FUTURE PILOTS CHECKLIST

By Captain Prat

Researched and written by Captain Prat

All rights reserved.

Without limiting the rights under the copyright reserved above, no part of this publication may be reproduced, stored in, or introduced into a retrieval system, or transmitted in any form or by any means (electronic, mechanical, photocopying, recording, or otherwise) without prior written permission.

For permission requests, please contact:

futurepilotschecklist@gmail.com

ISBN: 9798563766396

UPDATE 2025

Future Pilots Checklist

To you, our future pilot

Disclaimer

The views presented in this guidebook are solely the author based on available information at the time of writing. The purpose of this book is to inform readers not to provide professional advice. Readers are advised to research further and consult relevant professionals, such as flight training schools. Some information may be outdated. The author reserves the right not to update information. Readers are cautioned when acting on the information provided and assume all risks from such actions.

TABLE OF CONTENTS

TABLE OF CONTENTS 3
1 – INTRODUCTION 6
2 – IS FLYING FOR YOU? 10
    2.1 Reasons to become a pilot 10
    2.2 Age 13
    2.3 Being a woman 16
    2.4 Time to complete training 20
    2.5 Is it tough? 22
    2.6 Salary 24
3 – KNOW THE INDUSTRY 27
    3.1 A typical day as an airline pilot 27
    3.2 Phases of a flight 29
    3.3 Phases of training 30
4 – DIFFERENT PATHS AS A COMMERCIAL PILOT 32
5 – PROS AND CONS OF THE PROFESSION 44
    5.1 Pros 44
    5.2 Cons 46
    5.3 Suggestions 50
6 – ACADEMIC REQUIREMENTS 51
7 – SKILLS AND COMPETENCIES 56
8 – FINANCING TRAINING 70
9 – CHOOSING A SCHOOL 76
10 – OTHER PILOT SCHEMES 81

11 – MEDICAL CLASS 1 CERTIFICATE 86
11.1 Why is it needed? 86
11.2 Who does it, and who must have one? 88
11.3 What tests can I expect? 88
11.4 Failing the test 90
11.5 Time, cost, and validity 91
11.6 Limitations 91
11.7 Tips before the examination 92

12 – LANGUAGE BARRIER IN AVIATION SAFETY 94
12.1 The implementation of standard English in aviation 94
12.3 Levels 96
12.3 Suggestions 97

13 – TYPES OF WORLD LICENSES 99
13.1 EASA 100
13.1.1 EASA medical class 1 101
13.1.2 Licenses 108
13.1.3 Modular vs. integrated training 125
13.1.4 International students visa requirements 128
13.1.5 Tips to master your EASA training 129
13.2. FAA 130
13.2.1 FAA 1st class medical 133
13.2.2 Certificates 139
13.2.3 Visa requirements for international students 153
13.2.4 Tips to master your FAA training 160
13.3 EASA vs. FAA 161

13.4 Staying current 171
14 – CONVERTING LICENSES 174
    14.1 Foreign license conversion to FAA 174
    14.2 Foreign license conversion to EASA 175
    14.3 other conversions 177
15 – FLIGHT TRAINING 179
    15.1 Tips 179
    15.2 Equipment 186
16 – THE FUTURE OF AVIATION? 189
17 – LAST WORDS FROM ME 192
18 – YOUR FINAL CHECKLIST 195
19 - SUCCES STORIES 196
    19.1. Carlos´s age success story 196
    19.2. Joanne´s corporate success story 201
    19.3. Piotr´s training while working as flight attendant success story 205
    19.4. Andrea´s MPL success story 208
20- FREQUENT Q&A 213
21 - GLOSSARY 221
22 - ADDED DOCUMENTS SIDS, STARS, APP, TAXI DIAGRAMS 226

1 – INTRODUCTION

This book is a step-by-step guide designed to help you through the process of becoming a pilot.

I wrote this book intending to serve two significant purposes:

- ✈ to provide you with as much available information as possible about flight training, and
- ✈ to save you hours and hours of internet research on blogs, YouTube videos, and websites.

As a pilot, I've scoured through numerous books on the same topic written by fellow aviators. However, I was disappointed to find that they either focused solely on FAA training in the United States or EASA training in Europe. None of them covered both aspects.

As a European individual who took training under both regulations, I want to provide you with as much information as possible so you can decide what route is more suitable to you, regardless of where you come from. I will also discuss the conversions from your foreign license to EASA and FAA, which I have not found on any other guidebook.

In my opinion, it is essential and vital that you have all the options clear before choosing and deciding the next move in your pilot career. I have tried to make this guide as complete as possible. I am offering you exact steps and helpful tools,

In addition, the book's website provides extra supplemental information to satisfy your aviation curiosities.

Our world is full of acronyms such as FAA, ICAO, and JAR, etc. I will explain all these regulations in a friendly manner, since this book might be the first real contact with aviation that you have.

During my 10 years as a commercial Boeing 737 pilot, numerous motivated individuals sought me out for assistance and direction. Just as I did when I was a flight attendant, I reached out to other pilots I knew in the company and repeated the same questions. However, not all of you are in the aviation field and have the luxury of easily obtaining information from a pilot. Maybe you are a parent reading this book to help your son or daughter in their upbringing career.

There are many complexities and different paths to get to the flight deck. And no school or anyone can guarantee you a job position once you complete your training. Hence, it is good to have a clear picture of the entire process before you get started.

The two most renowned aviation regulations are the EASA (Europe) and the FAA (United States), described in detail in this guidebook, as well as how to convert your foreign license into them. You will also find a chapter comparing pilot training under both regulations.

My aim for this book is to address any questions you may have about this profession and provide guidance on how to navigate through flight training.

In conclusion, this book will help you to:

- ✈ Have an idea of the reasons why you want to become a pilot
- ✈ Clear up any industry misconceptions like age, gender, and background that stop many potential students from getting started
- ✈ Know how our world is: phases of flight, typical airline day, etc
- ✈ Know all the different jobs you can do as a pilot, not only work for a commercial airline
- ✈ Know the industry's pros and cons instead of believing in the unreal idea about being a pilot given by the media and tv shows
- ✈ Know what academic requirements are needed before joining flight training. You will be surprised!
- ✈ Know the interpersonal skills airlines look for in a professional, competent pilot
- ✈ Have different funding options and be able to combine them to finance pilot training
- ✈ Know how to look for the right flight school
- ✈ What to expect when acquiring the medical certificate
- ✈ Know the English level required and available tools to improve your aviation English
- ✈ Have a complete description of the two most worldwide recognized pilot licenses/certificates under the EASA and the FAA regulations
- ✈ Familiarize with procedure to convert your foreign license to the EASA or the FAA
- ✈ Great tips and tricks to excel during your training!

> ✏ By purchasing this guide, you get as a bonus 4 very inspiring success stories from colleagues who took different aviation paths. I hope they inspire and motivate you to pursue your career!

Now, get ready to take off and fly with me!

2 – IS FLYING FOR YOU?

Is this really for me?

What if I start training to realize afterwards, this is not what I had anticipated?

This chapter will dismantle many false beliefs and misconceptions about the industry given by old stereotypes in the media. We will go through the main reasons why someone decides to become a pilot, and if they resonate within you, you are on the right track! We will also cover topics like age, salary, or being a woman in the aviation industry, to the most asked questions such as if it is challenging and how long it takes to complete the training.

2.1 Reasons to become a pilot

Being a pilot is a highly vocational job; you must love, adore, and live for flying! Yes! Because it involves "sacrificing" weekends with your loved ones, missing special occasions like your kid's birthdays, or your best friend's wedding. It takes a significant financial toll on your bank account; hence it needs to be worth every cent and penny you will spend.

The reasons why I became a pilot, and anyone else who also decides to walk the same path are such as:

- ✈ It is a **rewarding** profession: you studied hard, you prepared very well, and finally, you get your wings. When a little kid during boarding approaches us, very shy but full of admiration and respect, it is gratifying to see how you created a forever memory by allowing them into the flight deck.

- ✈ You enter a **new world**: you will start looking at the world from another perspective; by this, you will work on a pressurized tube at 38.000 feet where the outside temperature is -66 degrees centigrade. You will have the most stunning views from above, cities, mountains, sunsets and sunrises, electrical storms, etc. You will also have to work during the night, when most people are sleeping. And on holidays and weekends, you will have days off while others are at work. Your life will be quite different from the rest of the people you know. You will spend many nights in hotels abroad. This "new world" might be difficult to comprehended and deal with by other people not in this industry.

- ✈ **Traveling**: if you enjoy traveling, this is a great reason to become a pilot! You will have lots of facilities to jump on an airplane and move freely around!

- ✈ **High demand** for the business: it is clear that more and more pilots are needed every year. You are about to enter a profession where there will be lots of doors open to all the different possible jobs you can do as a pilot (more about this on chapter 4).

✈ It is your **passion**: A burning fire that has fueled you since childhood. As long as you can remember, you have turned your attention to the sky. You know all airplane parts and watch all aviation documentaries. This job is pure adrenaline!

✈ **High profile**: some people enjoy the "status" the pilot job brings along with the financial income.

✈ You love **airplanes!** You have a profound admiration for these technical complex machines, their powerful engines, etc.

✈ **Challenging**: To become a pilot, you need to overcome many challenges, starting with financing your training, passing all your theoretical exams with a high score, passing all the practical flying exams on a first attempt if possible, maintaining currency afterwards. Most of us pilots like to excel and the challenge to become a better pilot today than we were yesterday.

✈ You are **learning new skills** like teamwork, leadership, communication, etc. You will become a skilled person flying and a great team player.

✈ You will broaden your **culture** since you will travel and most probably work surrounded by international people, you will expand your knowledge about many different cultures.

> ✏ I suggest you run through your mind your personal reasons why you think a pilot job is what you want.
> You might get this question asked at an interview!

2.2 Age

Am I too old?

Should I even start training?

What sort of jobs can I do if I am too old to join an airline?

This topic is the next big worrisome after finding the money to finance training. These are some of the questions many people have, and I will do my best to answer them in this chapter. One common misconception, or "false belief," is that if you are over the age of 30, you are too old to pursue a career as a pilot. Well, I'm here to challenge that limited mindset: there is no maximum age for becoming a pilot! Even if you are 80 years old, if you can obtain a medical certificate and get the license, you can fly! What you couldn't do is work and get paid to fly. Let's exlplore other options.

When I was doing my PPL course, we were around 5 "young" students, and there was a French gentleman in his 70's also obtaining his PPL license with us. I couldn't help but admire that fact. You can "decide" how young or old you can be. You are the one limiting yourself. Even though he was now past the retirement age of 65, he still pursued his dream of becoming a pilot by getting a PPL license. It is never too late to follow your passions and make them a reality, after all. As the saying goes, better late than never.

I started pilot training at the age of 27 and finished at the age of 29. I did my first line training flight with the company at the age of 29, just before turning 30, and I must say there were other cadets in my interview and type rating class, older than me.

I have a couple of colleagues who joined the airlines around the age of 40 and meet a few who joined being older. I remember from flight school, there were a few instructors after their 50´s and happily flying and teaching. They probably would have a hard time joining an airline. But, not for a corporate job or any other type of aviation job like crop dusting, banner towing, flying a medical airplane (air ambulance), or sightseeing tour flights!

Age will only limit where you want to head as a pilot. According to what I have found researching for this book, some carriers have a limiting age of 45 years and always require previous experience, but some do not have an age limit. If the airline does not limit the age requirement, do not be yourself the person to close the doors! Find your opportunity! You might get a surprise.

If you are around 35 to 40 years old, I would recommend doing the instructor course to expedite the experience (accumulate hours). The sooner you acquire 1500 hours, the sooner you will be an eligible candidate. You still have more than 20 years of flying if you start at this age!

> It is never too late to get started!

If you have reached the age of 65, it is no longer possible for you to work as a paid airline employee. You can only fly for recreational purposes. If that is your case, don´t be disappointed! You can still fulfil your dream of being a pilot and flying an airplane.

The following table show the average pilot´s age according to geographical location. As you noticed, the average is quite high.

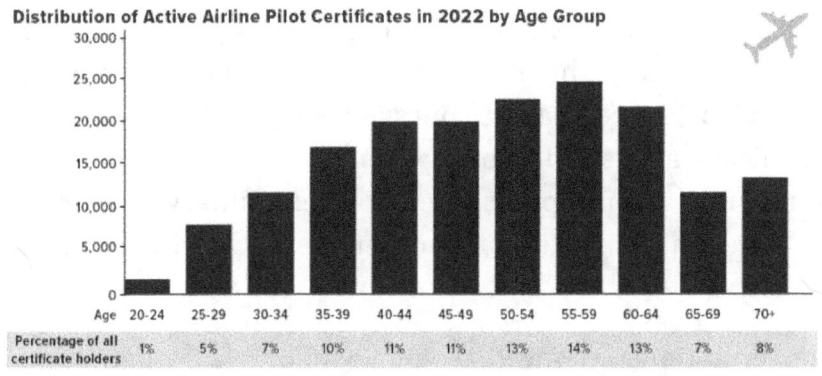

Source: usfunds.com

✏ Please read Carlos' inspiring success story on how he achieved his dream of becoming a pilot at the age of 40 on chapter 19.

2.3 Being a woman

As a woman, will I struggle in a male-dominated industry?

Will I be taken seriously?

My daughter wants to become a pilot. Will she suffer discrimination?

Lots of question marks. I, myself, being a woman pilot, can say I have never suffered any sort of discrimination or felt left apart, or that my professional opinion was not being considered or taken seriously. It is very much true that it is still a male-dominant world. The reason for this can be traced back to the highly technical nature of the job and the societal pressure that discouraged women from pursuing technical careers. Additionally, up until 30 years ago, most commercial airline pilots were former military personnel, further reinforcing the male-dominated atmosphere in the industry.

The first time I noticed a female pilot was when I was working for a regional airline In Spain, and, of course, there were only a few female pilots in that airline. Still, I clearly remember one of them telling me: *"I was also a flight attendant in this airline just like you,"* and now she was the

Captain. That was incredibly inspiring to me because, for the first time, I had connected the fact that a pilot's background didn't have to come from a non-technical job like a flight attendant. In my mind, I had this idea that pilots had to be engineers or mechanics or similar, regardless of gender. I also saw more female instructors and a few more female students during flight training. Now, I am happy to say that my current company and others have a significant number of female colleagues working alongside me.

Studies show that the percentage of women in aviation is around 5%. India is currently the country with the highest percentage of female pilots: 12,4%, according to the International Society of Women Airline Pilots (ISWAP). Followed by 12% in Finland, 8% in Sweden, 7,6% in France.

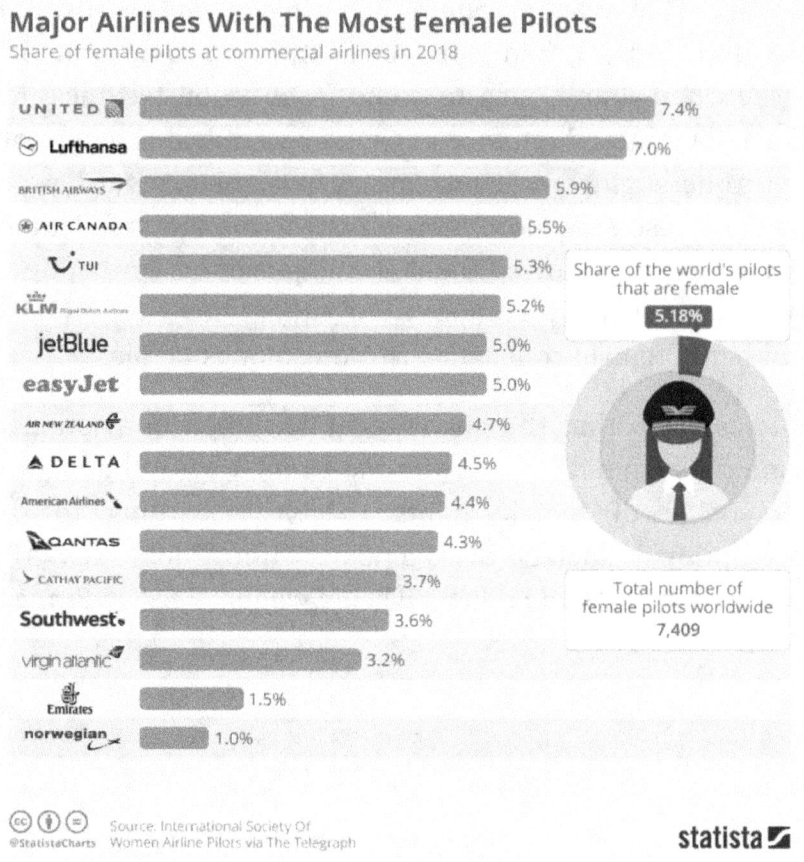

Source: Statista (Disregard that total number of female pilots worldwide)

Of course, being a female pilot is not the same in all countries. Like Saudi Arabia, some countries in the Middle East are just opening now the see the female figure as equal to a man, even if there is still a long way to go. The country saw its first certified female pilot in 2013, but she was unable to work in her own nation until 2019.

Let's discuss the issue of pay discrepancy. In my company and others throughout Europe, men and women are paid equally, regardless of gender. Pilots earn their salary based on the number of hours they fly; in this aspect, women do the same job as men in the cockpit. However, if a female pilot becomes pregnant, she will not be able to fly and therefore her earnings will be affected. Considering that pregnancy lasts nine months plus additional time for breastfeeding, it can amount to a significant period where she is not earning the same as a male colleague who only takes a few months for paternal leave In the same situation.

Finally, there is a big surprise element when people see a female pilot. Usually, in certain countries, people are still in disbelief to see a woman at the jet's controls. On the contrary, I have never perceived any adverse reaction from a passenger or colleague. While Spain may still be lagging behind in some areas, the only negative issue I have encountered is a few distasteful jokes from individuals outside of the aviation industry. However, I have learned not to let these comments affect me, as they stem from ignorance and I dare to say, even envy.

All women out there and parents of any young girl who dreams of becoming a pilot, it is a safe world for women, and it looks like it will be much more common in the future to see women at the controls of an airliner!

> ✏ **NOTE**: You can find a blog post on the books website with a list of female pilot scholarships.

2.4 Time to complete training

How long is the training?

The answer to such question will depend a lot on yourself and the school. Let me explain this in detail: the typical professional pilot training takes a minimum of one and a half years, 18 months, some people manage to squeeze it all in one year! If you are in a hurry to become a pilot and you can allow yourself financially to be a full-time student, then you can do it! Be aware that this means committing 100% to the training.

If you are doing an EASA training, I advise you to do an integrated course since it is faster than modular. Keep in mind the disadvantages of a integrated training program, which will be discussed in more depth in chapter 13.1.3.

Maybe you are working meanwhile you are training, perhaps you are taking your time in between modules because you are not in any hurry. Whatever it is your decision, it is the correct one, there is no right or wrong. Remember that airline recruiters will look into your resume to see how many exams you failed. If the number is high, this could reflect on an irresponsible/immature personality, such as skipping classes and partying.

Ultimately, it will be up to your commitment to flight training and the financial and time constraints set by your parents (if they are the ones footing the bill).

Another possibility for reducing the pace of your training could come from the school. I started my PPL training in Barcelona at a local flight school that was more sort of an aeroclub than a professional school back then, where I spend one year doing the PPL. In order to take the ground exams, I had to make a few trips to Madrid because they were not offering them in Barcelona, adding extra money and time to my training. After PPL, I decided to continue studying in the United States. I had found this FAA-EASA school in Florida where I could expedite my 14 ATPL exams in only six months, which was great! And I only had to drive two hours to do them instead of flying to another city. I spend the remaining of that academic year doing hour building and the multi-engine instruments rating. In the next and final year, I completed the commercial license and converted from FAA to EASA the ME-IR and, finishing everything with the MCC training. During those three years I did my training, I did not have any holidays at all, even after the first year I started working as a pilot. A total of four years focused on working towards completing pilot training and getting my first job! I could have probably reduced this to 2 years if I had done all the pilot training in the same school.

I have colleagues who were victims of flight schools bankruptcy during their training. These students lost their money already paid to the school and were unable to finish obtaining their pilot's license. That will rarely happen to you, but it would be another reason why a school could delay your flight training.

If you only intend to do a private pilot for fun, it might be OK to go to a flight club and take some time to do it. But if you aim towards professional training, I strongly advise you to look for a school where you can do everything from the beginning on a fast track, and where ground exams are taken nearby or in the same school.

More about choosing the right school in chapter 9.

2.5 Is it tough?

Is becoming a pilot too challenging?

I struggle with math.

Do you need to be exceptionally intelligent to become a pilot?

It is insane the number of people that believe flying an airplane is too complicated. I get asked so many times this question. There is this misconception that the training is tough. I can guarantee you that an engineering degree is much more difficult. The maths and physics applications are quite basic; we are constantly doing fuel and time calculations on the day-to-day operation. Nowadays, we have everything digitally, an app that will do all the calculations for us, making the operation safer and faster. Previously, you had to double-check the manually done calculations and spend time with paper tables interpolating values.

The bigger the airplane Is, the more systems It will have, and the more complex It will be. But let me share some advice: it all comes down to practice and preparation. If you can learn to drive a car, you can surely learn to fly a small airplane. When you progress to a turboprop or a jet, there will be new systems that you will have to learn. Everything will be practiced on a simulator. In case of emergency, there is a checklist to follow.

Now, ground school. Many people are worried about failing knowledge tests and not having a previous maths and physics background. I can tell you some topics are difficult to understand and quite complex, but nothing you cannot master with a lot of study, practice, and the will to get through it. Many of the topics you will cover in your ATP or ATPL courses may never be applicable or relevant to your future career as a pilot. It is worth attempting. During my time at flight school, I observed some classmates who had no comprehension of the questions being asked. They simply memorized the question and the answer without truly understanding the concepts behind. Yet, to my surprise, some of them turned out to be incredibly skilled pilots. However, I do not recommend this approach. Instead,

I suggest seeking clarification from your peers if you struggle with comprehending any material.

A crucial principle In aviation Is preparation, as It helps make things less difficult and complicated. It's always important to stay ahead of the game. Preparation is key for any task or challenge. Before taking an exam, thoroughly study and prepare using the question bank to assess your readiness. If you are not adequately prepared, the school will not allow you to present at an official exam. Before you go for a flight lesson, prepare carefully at home, and repeatedly review the procedures in your mind. It is a way to make training easier for you. One friend of mine told me: *"use the shower time to recite the procedures!"*

Returning to school after a long break can be difficult, especially when it comes to getting back into the habit of studying. Don't be too tough on yourself and take it one step at a time! One step at a time. We have all been there. I started pilot training at the age of 27, nine years after I had finished school!

2.6 Salary

How much do pilots make?

Are pilots wealthy?

One common misconception is that all pilots make a lot of money. However, the salary greatly varies depending on the airline and type of aircraft flown. In this chapter, we will explore the average salaries for different pilot positions in various countries.

For most airlines, it is typical for your hourly wage to increase annually as you gain more experience. This would indicate that your income will not remain constant after a few years of working for the airline. Seniority is valued and paid accordingly.

The maximum salary earned by a captain is around 300.000$ a year (25.000$ a month). Some airlines in China pay those amounts, as well as some cargo companies in the U.S. The minimum salary is around 20.000$ (1600$ a month) for jobs like flight school instructors or photography pilot. But the average, according to *www.salary.com* in the United States, is of 137.000$.

AIRCRAFT TYPE	CAPTAIN	FIRST OFFICER
LARGE JET (more than 12.500 pounds)	$288,338	$157,343
SMALL JET (less than 12.500 pounds)	$220,000	$135,345

These amounts are gross

According to the information I found, in England, for short-haul first officer, the salary range is between 35.000 to 60.000 pounds, and for the long-haul is between 45.000 to 120.000 pounds. As per a Captain, the short-haul range is between 60.000 to 100.000 pounds and long-haul between 80.000 to 170.000 pounds. The UK average salary of a pilot is 72.000 pounds. Europe's averages are between 27.000 to 102.000 euros for first officers. And for a Captain between 72.000 to 144.000 euros yearly.

The following table lies the <u>maximum annual salary a Captain</u> can make in different world regions:

COUNTRY	GROSS SALARY DOLLARS
CHINA	243,000
UNITED STATES	375,472
CANADA	94,900
EUROPE	270,000
MIDDLE EAST	178,335
INDIA	177,271
JAPAN	76,884
SOUTH AFRICA	2,525

Source: AI search engine

3 – KNOW THE INDUSTRY

This upcoming chapter will serve as a brief introduction to the industry you are about to enter. Here, you will familiarize yourself with key acronyms and topics that will be referenced throughout the rest of this book. Additionally, this section will provide insight into what a typical workday may look like and what you can expect once you have completed your training and obtained all necessary licenses for the job application process.

3.1 A typical day as an airline pilot

The nature of your life as a commercial pilot will greatly differ based on the aircraft you operate and the specific operations of your company.

A typical day usually starts by reporting 1 hour before departure to the airport's airline operations office or, even at the airplane. Together with the rest of the crew, you start looking at the paperwork together with your flight deck colleague/s: check the weather, en-route charts, aircraft technical status, any airport delays, and flight plan details such as the number of passengers, flight time, flight levels, etc.) You and the team will jointly determine the final fuel requirements and any additional needs for that day, which may include coordinating with maintenance if the aircraft has technical issues. While the Captain has the ultimate say, the decision will be made through team discussion.

Afterward, you will meet with the flight attendants. Usually, the flight purser (the flight attendant in charge) introduces him or herself to the rest of the crew and discusses the airplane's positions assigned to the rest of the team.

The people who you fly with is always different. The scheduling department alternates the crews, so you don't fly with the same people every day, which adds excitement and enjoyment to the job. Usually, it's the team you're with that makes the day special!

Pilots brief the flight attendants on details like the flight duration, expected turbulence, weather at the destination, etc. Once at the aircraft, there are many things to deal with: the refueling agent, tech log (An aircraft technical manual where we can review in detail the most recent maintenance performed on the aircraft, as well as any technical issues encountered by the previous flight crew), the ramp agent who coordinates the boarding and brings us the final load document, etc. Once all the checks are completed, we brief the departure chart (SID) and run the checklist. Simultaneously, the flight attendants are doing the boarding and preparing the cabin to ensure an on-time departure.

Once we are ready, pilots request ATC's clearance to push back from the gate, start the engines, and taxi to the runway in service.

And off you go for your flight.

3.2 Phases of a flight

Here, you will gain a clear understanding of the typical phases of an airline flight from point A to B following pushback and engine start.

1- Pre-departure (covered in point 3.1)
2- Taxi from gate to the runway
3- Take off and SID (Standard Instrument Departure).
4- Climb phase to the TOC (top of climb)
5- Cruise phase till TOD (top of descent)
6- Initial descent
7- STAR (Standard Instrument Arrival)
8- Approach and landing
9- Taxi from the runway to the gate
10- Post-flight

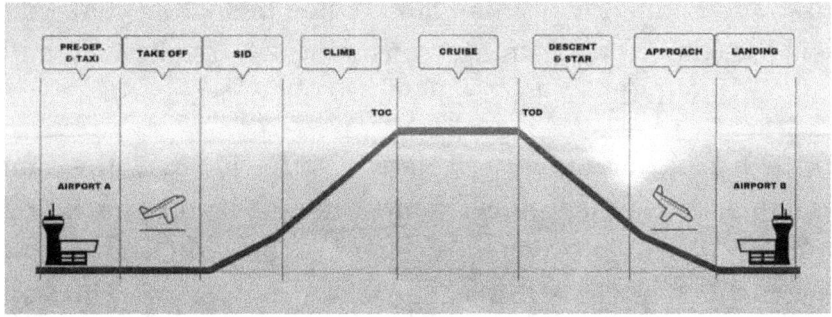

After your post-flight duties, either you go home or to the hotel together with all the crew. If you go to a hotel, you will have time off to rest till your next shift starts.

> ✏ **NOTE** If you want to see how a SID, START, Taxi diagrams, and Approach charts look like, you can find examples of them in chapter 21

3.3 Phases of training

Outlined below are the various training stages you'll encounter after obtaining your licenses, so as you proceed to the following chapters, you'll have a better understanding of the terminology used.

For EASA / Europe:

Once you start a type rating, this consists of ground school first and simulator training after. Once this is finished, you will do the base training, consisting of 6 take-offs and landings with the real aircraft empty (no passengers).

Once base training is completed, you will start the line training. It's called line training because you are doing regular line flights with passengers. You will fly with an instructor Captain and a safety pilot until you are checked safe to land in case of Captain incapacitation. After you get rid of the safety pilot, you will continue flying few more sectors until you are fully line checked on all company procedures and the airplane's safe operation. Now you are finally a line first officer!

For the FAA / United States:

There is a slightly different procedure under the FAA regulations. Pilots are not "usually" hired by a company until reaching 1500 hours, following the usual path to the flight deck by first becoming a flight instructor or, do time building.

Afterward, the regional airlines are the ones recruiting you without jet time. You will spend a few years at a regional airline and possibly become a Captain there, to later on, jump to the big ones like AA, Delta, or United, where you will go back to being a first officer for the next 10-15 years. Until you become a Captain again, and even longer if you jump to the long-haul.

After completing the type rating (ground school plus simulator sessions), you will start your Initial Operating Experience (IOE) directly with the real aircraft and passengers. You will still be flying around 20 to 25 hours with a Line Check Airman (LCA) that will be teaching and guiding you. The last leg (flight) will be a check ride; if you pass it, you are already a line first officer!

4 – DIFFERENT PATHS AS A COMMERCIAL PILOT

During my time as a flight attendant, I worked for a few different airlines. One was a Spanish regional airline with small airplanes flying mostly domestic; the other one was the biggest company in the Middle East flying worldwide transatlantic flights. Naturally, I was a flight attendant, but my lifestyle closely resembled that of the pilots. We all reported at the same time and stayed in the same hotels.

Having explored both sides of aviation, I can assure you that not all of them will be appropriate for your situation at a given time. I want you to know that you have many options, and with time, you can move to another company that offers you something different, that aligns better with your current circumstances. Maybe you are looking for more stability? Or perhaps you got bored and need more excitement?

Some airlines will require you to spend 15 to 20 days a month away from home in hotels; on some others, like my current airline, you will sleep every night at home. Others, such as corporate flying, will be on call 24/7 (almost) and never be able to make short-term plans. For some other jobs such as flight instructor, you can mostly tailor the schedule to suit your preferences. If you become a firefighter pilot, you will work only summers, etc. The aviation world offers a wide variety of jobs and lifestyles that I am describing in detail in this chapter.

At the start of your career, 90% of the time, you will need to seize any opportunity that comes your way since you lack prior experience. Even if it involves flying a propeller plane instead of a jet initially, working as a bush pilot, or becoming a flight school instructor for a few years, even if those roles weren't part of your original plan. Aviation is volatile. It goes with the current economic situation. In some periods, there might be a high demand for pilots. Others, like post 911 and Covid19, might be an excess of pilots not needed for a while.

We usually say the aviation world is small, but that depends on where you move. Most people equate being a pilot with a standard airline pilot. But the are many options, and I want you to be aware of them. You can enjoy flying in many different ways, and with that, your lifestyle can vary so much from one another.

Being a bush pilot is definitely not the same as being an airline pilot!

Here are some of the different pilot jobs described:

✈ Main airline

The largest airlines in each nation primarily focus on passenger transportation and have stopovers in major cities  across the globe. Pilots spend about ten nights a month in hotels, and they work with a new crew on every flight. Depending on the aircraft type, the team might be from 6 to 18 various members, making the stopovers and work environment very dynamic.

Some examples are Iberia, British Airways, Emirates, Lufthansa, American Airlines, Singapore Airlines, Air Nippon Airways, to mention a few.

✈ Corporate / executive aviation

They are employed by either the aircraft owner or a company offering private flight services. This is the pinnacle of aviation, as they transport celebrities and royalty. The schedule is unpredictable because it revolves around the client's needs. The team remains consistent and small, often comprising the same Captain

for months and just one flight attendant. It's a unique experience where you meet fascinating individuals while flying corporate!

Some examples are: NetJets, VistaJet

✈ Airline instructors both ground and air

This role typically requires extensive experience with the specific aircraft type. With students often asking numerous questions, it's a great chance to enhance one's knowledge and broaden experience. They offer line training to cadets and conduct line checks for the regular crew.

✈ Simulator instructor

The role involves instructing students on aircraft type ratings. These instructors are employed by simulator centers such as CAE or Pan Am, or by airlines. Major airlines typically have their own simulators for training their staff. Certain companies permit their simulator instructors to simultaneously maintain their flying skills by allowing them to fly, ensuring they stay current.

- ✈ **Flight school instructor both ground and air**

A great chance also to expand knowledge. Becoming an instructor on your flight school will allow you to build many hours quickly and gain experience already teaching. It is quite a common practice in the U.S. after completing your training to become an instructor at the same school.

- ✈ **Bush pilot (small rural communities in Asia or Africa)**

Bush pilots often operate in isolated rural areas of Africa or Asia. For instance, remote communities in Papua New Guinea might require medical supplies or need to transport an injured person to a hospital or a major city for various reasons.

This type of work typically involves flying single-pilot

aircraft. The destinations often lack proper asphalt runways and don't have instrument approaches. Many pilots find this style of flying to be quite enjoyable!

>   Some examples are Susi Air, Wilderness Air, Westair, AirTech Global, Ulendo Airlink, and Solenta Aviation.

✈ Seaplane pilot

Seaplane pilots operate over lakes and tranquil water locations such as the Maldives. Their primary job is to transport passengers traveling to resorts or those with homes on secluded lakes that are difficult to access by road. Seaplanes are approved to fly solely under Visual Flight Rules (VFR) conditions.

Some seaplanes are considered amphibious; these airplanes are certified to land on both ground and water. For example: when taking passengers from the mainland airport and transferred to the lake or islands.

>   Some examples are Trans Maldivian Airways, Kenmore Air, Harbour Air Seaplanes or, European Coastal Airlines.

✈ Agricultural pilot

Agricultural pilots typically operate solo, as the planes used for this work are single-engine (or single pilot?). These pilots are highly skilled because they fly at low altitudes to spray the crops, maneuvering near power lines. This work is seasonal, as crop dusting isn't required throughout the year.

✈ Cargo freight pilot

Cargo pilots typically fly goods for major companies like FedEx or Amazon, primarily during nighttime hours. It's a well-compensated role. The crew is generally made up of 2 or 3 pilots, depending on the size of the aircraft. The schedule may not be rotational, so you could spend ten straight days traveling to various locations.

Some examples are Cargolux, DHL, Ups, FedEx or Jade Cargo.

✈ Regional/low-cost airline pilot

Regional pilots typically operate flights that are either domestic or confined to a specific area, such as all of Europe.

 They fly smaller passenger jets like the CRJ or the Embraer family and medium-range airplanes like the B737 or A320. Depending on the airline, they might sleep home every night or stay at local hotels.

> Some examples are Air Nostrum, Sky West, United Express, JetBlue, CityJet, or Olympic Air.

✈ Charter pilot

Charter pilots often work on a seasonal basis; particularly when there's a huge demand for flights to popular tourist spots during the summer. Charter airlines handle these flights. Additionally, companies might hire a charter flight to transport large groups, such as football teams. This option serves as a cost-effective alternative to private flying, tailored for large groups and addressing market needs in specific situations.

> Some examples are Sun Express, Atlas Air, Smart Wings, or Air Transat.

✈ Military pilot

Military pilots fall into the high-risk category. They typically  undergo state-sponsored training rather than attending a standard flight school and paying for their education. Their duties can range from transporting goods, soldiers, and explosive materials to escorting dangerous criminals or prisoners, or conducting rescue missions in foreign countries. They possess strong discipline and have a distinct working style compared to other pilots.

✈ Government pilot

Primarily composed of ex-military personnel, this group understands the importance of secrecy and discretion in their missions, which often involve transporting politicians and high-ranking officials, such as head of states. The work is on-demand, requiring them to be prepared at all times. These flights are typically accompanied by military fighter jets for security.

✈ Emergency pilot

Certain emergency pilots serve as coast guards or operate

air ambulances, which are small jets equipped with medical gear. While helicopter pilots often perform these roles, others, such as firefighter pilots, operate planes fitted with large water tanks.

They work seasonally from May to October, primarily in Europe. After that, some pilots head to South American countries, such as Chile, to work during their summer season for the remainder of the year.

✈ Test pilot

Whenever companies like Airbus or Boeing develop a new aircraft model, test pilots conduct experimental flights to check each aircraft system thoroughly before it can be certified for safety and sold commercially. Additionally, they assist in air crash investigations by replicating the flight conditions on a simulator or, if possible, on the actual airplane.

✈ Space pilot

It is the ultimate level of flying, in my opinion. I had to include a space pilot in the list of pilot jobs. Were you aware that all astronauts landing the space shuttle were all pilots? NASA's requirements to apply for an astronaut job are to have a science degree, pilot experience and a U.S. citizenship (which is a drawback for the rest of us).

You never know how high your aspirations can take you!

As you can see, there are numerous pilot opportunities available! The realm of piloting extends beyond just airline positions, contrary to popular belief. I've provided a glimpse into the wide variety of jobs a pilot can pursue. Remember, some of these roles are excellent alternatives, especially if you've finished your pilot training during an economic downturn, such as during Covid-19, when major airlines aren't hiring.

> ✐ **NOTE**: You will **not** find the helicopter job in this section since it englobes an entirely different license/certificate. You cannot fly helicopters with an airplane license/ certificate.

✏ Please read Joanne's success story flying corporate on chapter 19.

5 – PROS AND CONS OF THE PROFESSION

There are numerous misguided reasons why individuals choose to become pilots. The list of pros and cons provided here is intended to help you determine whether the advantages excite you more or if the disadvantages are significant enough to deter you from starting this journey.

5.1 Pros

✈ Fulfilling profession

A saying goes: *"Do what you love, and you will never work a day in your life."* If flying is what you like, you will feel very fulfilled,

and you won't see it as work because you'll be earning money by doing what you love the most.

✈ Traveling and exploring cities and countries

Being a commercial pilot makes traveling more convenient. With discounted tickets, you can easily hop on a plane and explore new cities with friends and family. In fact, everyone will be asking you to get them flight tickets!

✈ Great salary

I must say the salary improves with seniority, experience, and the type of aircraft you fly. In the beginning, it might not be great, but some pilots earn up to 500.000$ a year! (ask ChatGPT)

✈ High demand when the market is growing

Each day, more individuals are traveling due to the availability of low-cost airlines. Companies are establishing new routes between cities and acquiring additional aircraft. Even if the market experiences a downturn, it is likely that it will bounce back eventually, leading to a renewed demand for pilots.

✈ Exciting

This job is packed with adrenaline! If you love the thrill of lifting a heavy complex machine from the ground into the sky, breaking through the clouds, watch the most amazing sunsets and sunrises, even watch the storms from far away in the sky, this is for you! I can guarantee you two days are never the same!

✈ No work carryovers

This profession allows you to leave your work at the airplane after you land. This means you can fully enjoy your free time and disconnect!

✈ Airline benefits and discounts

Major airlines offer numerous perks to their staff, such as discounts at gyms, dinning, flight tickets, car rentals, and more.

✈ Free time

You will have granted at least ten days off a month, sometimes even half a month. For instance, with my airline, I enjoy four consecutive days off! This allows me plenty of time to pursue hobbies, travel to nearby countries, or even attend classes.

✈ Having a family

Most airline benefits extend to family members, such as discounted tickets and accommodation. In some companies, the airline may also cover family medical insurance and children's international school tuition. Pilots are entitled to maternity or paternity leave when they have a new baby.

5.2 Cons

✈ Training is costly

This is likely the most expensive option on the list. You'll require approximately 80,000 to 110,000 dollars or euros. If you don't have the funds, you'll need to earn and find a way to make it a happen.

✈ Working holidays and weekends

Indeed, working means that Sundays and New Year's Eve lose their significance. You'll likely have days off when your friends and family are busy at work, and you'll find yourself working on weekends and holidays when they're free. This isn't always the case, but it happens frequently.

✈ Being away from your country, culture, etc.

Sometimes, securing a job in aviation may sometimes require relocating to another country. Adapting to a new culture and language while working and living there can be quite a cultural shock. However, some individuals view this as an advantage.

✈ Getting the first job without experience.

Beginnings are not easy. Some airlines are looking for experienced pilots (high hours and type-rated), but where do you start? You might have to begin instructing in your flight school or take that crop dusting job before you can gain more experience to apply for another position.

✈ Fitness for a medical certificate

Undergoing the annual medical fitness test can be nerve-wracking. What if the doctors discover something that grounds me from flying? Or what if I fail the vision test and must start wearing glasses? These are some of the concerns that run through one's mind before undergoing medical examination.

✈ Jet lag/lack of sleep/fatigue

Flying at night and covering long distances across multiple time zones can leave you feeling particularly exhausted and occasionally disoriented. These disruptions may cause you to not enjoy the job at times.

✈ Volatile industry

You've finished your training and have begun job hunting, only to find the market plummeting due to events like 9/11, the Boeing Max issues, and Covid-19, causing airlines to halt hiring. What should you do? It might take a couple of years to secure a pilot position. Even if you do land a job, there's a risk your airline could potentially go bankrupt, forcing you to relocate across the globe for work. The more stable the airline, the better its chances of weathering an economic slump.

✈ Irregular schedule

Contrary to your days off, when you work, you work extensively, it can extend to 12 consecutive hours during peak periods.

On early shift days, you wake up as early as 3 a.m. Your schedule can vary significantly, switching from night shifts to early mornings, and then to evening shifts. This lack of a consistent routine is definitely not ideal for your body.

✈ Stress and tension

When the sun is bright and the sky is clear, everything feels ideal. However, not every day goes this smoothly. On some days, pilots face difficult weather conditions, along with delays and technical issues. It's crucial to manage stress by tackling one issue at a time. Additionally, the long hours spent sitting can lead to back tension and discomfort.

✈ Cosmic radiation

Airlines keep track of the cosmic radiation exposure of their pilots and cabin crew, restricting the number of hours they can fly annually. There's no need for concern, as the radiation levels are monitored and deemed safe. Pilots on ultra-long-haul flights receive more cosmic radiation. If this worries you, consider pursuing an aviation career that doesn't involve flying at high altitudes such as tour pilot, agricultural, parachute pilot, or working for companies that use turbo-props planes as they don't fly as high as jets.

> ✏ I recommend choosing the points that stand out to you from each list of pros and cons and creating your own list below. Check which of your lists is longer; the longer one can guide your decision.

I trust this chapter assists you in determining which commercial pilot role suits you best and how to manage its drawbacks. Personally, the most challenging aspects are nighttime shifts and the jet lag from crossing time zones. However, some individuals don't mind night work and may even favor it. If you have a family and children, you might not appreciate a corporate aviation position as much as someone who is single and eager to travel globally.

5.3 Suggestions

Here are some suggestions to get a feel for flying before signing up for a professional program or a private pilot course: Visit a flying club or school and request a familiarization flight. This typically involves spending an hour in a small training aircraft with an instructor who will fly the plane and might even allow you to make a few gentle turns. Pay attention to how you feel during this experience. Are you thrilled by the view? Do you crave more of it? Does the experience excite you? Are you comfortable being at altitude in a small aircraft?

These elements are crucial in deciding whether this is something you will enjoy and want to learn. For me, just visiting the flight school was incredibly thrilling! I was eager to begin and have my fresh new books and flight bag with me!

Another idea is to visit a large aircraft simulator, such as a B737 or A320, and observe pilots or trainees as they operate it. This experience closely resembles that of an actual airline pilot. Are you intrigued by every switch and system? Do you have countless questions for the pilots about the systems, procedures, and operations? If this describes you, then you're similar to me, and the good news is that you're on the right track! You can do both: attend a simulator session and a familiarization flight.

6 – ACADEMIC REQUIREMENTS

Do I need a university degree?

Do I have to be good at maths and physics?

Do I need an adequate level of English language?

You don't need a degree or specialized training to join a flying academy or become a certified pilot with an airline. Most flight schools only require a high school diploma and that you are at least 18 years old. Some might also request an English proficiency certificate. However, some airlines may favor hiring a pilot with a university degree over one who doesn't have it. It's important to stress that this isn't a universal rule, and it doesn't determine a pilot's competence or skill level.

Certain flight schools now require prospective students to take an entry-level test, particularly if they lack prior knowledge in math and physics. This trend has been increasing; in 2013, only a handful of large schools conducted such assessments. The same requirement applies to proficiency in the English language.

There's no reason to worry just yet. You can find a wide range of online pilot aptitude tests and preparation resources available.

Future Pilots Checklist

I want you to understand that there are countless opportunities available to you; don't shut the doors on your dreams. If you lack a science background, you can still pursue one. You can sign up for one of these training programs or practice on your own to develop skills in math and physics.

The same goes for learning English. If you're not fluent, your initial step should be to spend time in an English-speaking country to improve your fluency. This is essential since all your training, including theory exams, instructor lessons, and ATC communications, will likely be conducted in English.

Here are some examples of the fundamental math and calculations you'll encounter in your private pilot training. The course involves a good deal of calculations using mathematical and physics formulas, along with the analysis of tables and graphs.

Once you are on the line, things get easier because most airlines equip their pilots with a tablet. This device uses software to handle all the data and calculations, such as speeds, landing distances, and brake cooling, which saves both time and reduces potential errors.

One of the daily calculations pilots do is transfer the fuel from liters to kilograms (as used in Europe) to ensure that all aircraft weights are consistently in the same unit. We input all the information into our performance app to verify if the airplane is operating within its envelope limits. *

Here's an example of an aircraft within limits and one that is not. Can you identify which one is correct?

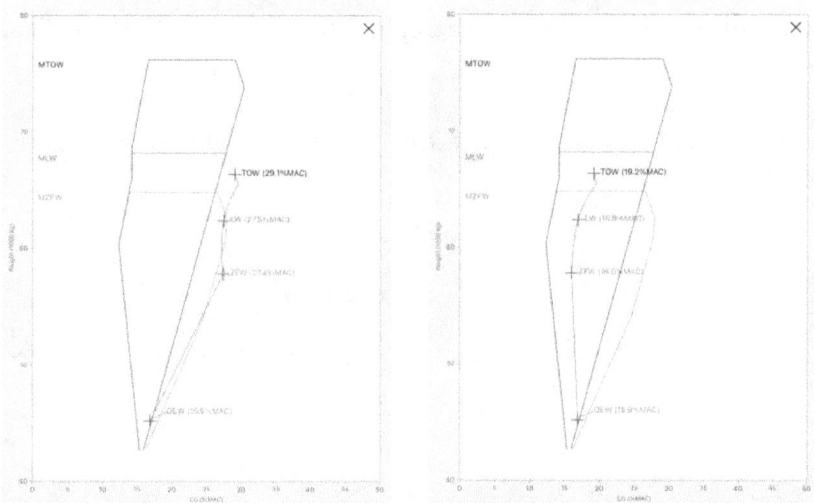

*Boeing performance app: left photo is outside the envelope, out of limits.

TOW (Take Off Weight) is 29.1% whereas the right TOW is 19.2%

The right one is correct, within limits.

*SERVICE ENVELOPE

The service envelope of an airplane refers to the range of operational limits within which the aircraft can safely and effectively perform. It includes parameters such as airspeed, altitude, load factor, and structural capabilities, as well as the allowable range for the center of gravity (CG). Maintaining the CG within specified limits is critical for stability and control, ensuring the aircraft operates safely throughout its performance boundaries.

A basic private pilot calculation to find out time:

$$T = \frac{D}{GS}$$

D= 250nm
GS=150 kts
T = ?

The time to fly 250 NM at a GS of 150 knots is 250 ÷ 150 or 1.66 hours. (The 0.66 hour multiplied by 60 minutes equals 40 minutes.)

Answer: 1h and 40 min

You can turn around the formula to find out the ground speed (GS) or the distance (D).

To conclude this chapter, here are some motivational words: don't lose hope if you encounter challenges. Keep practicing until you gain confidence. Many tasks you face in the theory exams won't need to be repeated. I've seen colleagues fail half of the 14 ATPL exams and still succeed in passing them, eventually securing jobs with airlines.

7– SKILLS AND COMPETENCIES

What qualities define an ideal airline candidate?

What particular skills or personality traits do airlines seek?

What is required to become a proficient and skilled pilot?

The priority is always to fly from point A to B safely and efficiently. To achieve this, airlines strive for competent professionals who can work together, regardless of culture, religion, or different backgrounds, even when they have never met each other. It is a complex and challenging environment. The following set of skills set the operation for a successful outcome in any abnormal or emergency situation when the pilot applies them. Sometimes, there will be no right or wrong way of solving a problem. Still, a decision has to be made in the most efficient and safe way, involving a combination of experience, knowledge, procedures, and the rest of the skills mentioned below.

According to a few investigations and studies, the results that make a very successful airline candidate are a mix of:

- ✈ Aircraft handling skills (25%)
- ✈ Technical knowledge and aircraft operation (25%)
- ✈ Competencies (non-technical skills) (50%)

Creating an environment where people can communicate and engage freely is crucial. Crew Resource Management

(CRM) was established in the 1980s to minimize human errors that lead to accidents. It emphasizes interpersonal skills in the cockpit, like communication, teamwork, and decision-making, among others, which are detailed below with examples from actual fatal accidents. By the 1990s, implementing and training crews, including flight attendants, in CRM had become a common standard.

> **CRM:** Crew Resource Management is the effective use of all available resources for flight crew personnel to assure a safe and efficient operation, reducing error, avoiding stress and increasing efficiency.

✈ Communication

The process of passing information and understanding from one person to another.

Miscommunication has caused numerous airplane disasters since the early 1970s. The most notable instance is the 1977 collision of two Boeing 747s at Los Rodeos airport in Tenerife. Its significance cannot be overstated.

Effective communication is not one way; it's a two-way process, where instructions need to be acknowledged and read back in a clear, concise, and precise manner.

Communication can be a barrier to the people you work with, especially if they come from different cultures. We fly worldwide around the globe. And that is why airlines require pilots to have an ELP (English Language Proficiency). More about this topic in chapter 13, where I give great tips to get better at aviation English.

Other barriers to communication could be the noisy cockpit (single-engine or open cockpit), high workload and stress, fatigue, as well as interruptions or distractions.

Pilots and controllers use a common standard aviation phraseology. An example of not using standard phraseology is Avianca flight 52 that crashed on a hillside in New York in January 1990. The crew did not declare an emergency (Mayday, Mayday, or PanPan) when running low on fuel, thus not getting landing priority.

✈ Application of procedures

Procedures: a series of actions conducted in a particular order or manner.

The application of procedures is meant to help pilots during challenging and high workload situations. Each airline and aircraft manufacturers have designed a set of procedures to follow in strict order, what we call the SOP's (Standard Operating Procedures). These procedures, along with checklists, are usually developed by flight instructors and engineers.

There are instances when a particular checklist may not be appropriate for a certain scenario, such as dealing with a gear malfunction amidst bad weather. In such cases, pilots should focus on navigating the aircraft out of the hazardous weather before addressing the gear malfunction checklist.

Turkish Airlines flight 634 in January 2003 failed to stick to SOP's to start a missed approach when being not visual at minimums. Instead, the flight continued descending without visual of the runway until crashing on the ground, killing 75 out of 80 occupants.

✈ Leadership & teamwork

Leadership: the art of motivating a group of people to act toward achieving a common goal

What differentiates a boss from a leader? Generally speaking, leaders lead by example, and bosses tell you what to do. A good leader is expected to be willing to listen to others, think outside the box, and have the ability to adapt and perform under abnormal circumstances. A great leader keeps on learning from others constantly. He or she recognizes that team members have different strengths and limitations but ensures good communication, respect and provides purpose, so that individuals are unified to work together to achieve the desired result inside a non-punishing culture. A great leader considers all crew members, delegates tasks, and provides constructive feedback to resolve conflicts.

Experience, when paired with training and knowledge, shapes a Captain into a great leader from whom others can learn.

Teamwork: a group of people with matching skills working together toward a common goal or purpose.

It occurs when every team member (ground and air) performs and contributes in the best possible way to achieve a safe and successful operation. An example of poor teamwork and poor leadership is reflected on Air France flight 447, where two first officers were left at the flight deck, not knowing who had control of the aircraft. This resulted in crashing into the Atlantic Ocean, killing all on board.

On the contrary, a great example of teamwork is US Airways flight 1549 in January 2009, "successful" ditching into the Hudson River by Captain Sully saving everyone's life.

✈ Problem-solving & decision making

Problem-solving: the act of determining the cause of a problem, identifying, prioritizing, selecting alternatives for a solution, and implementing them.

When encountering a technical issue, a pilot relies on their expertise and available data to devise an effective resolution. The ever-evolving airline industry, with its schedule disruptions, delays, technical glitches, and rapidly changing weather conditions, requires flexibility, adaptability, and swift problem-solving. Pilots often use acronyms in aviation to help recall crucial parameters.

The PAVE model for pre-flight planning:

P= Pilot-in-Command

A= Aircraft

V= Environment

E= External Pressures

And the **CRAFT** model used to read back IFR clearances:

C= Clearance Limit

R= Route

A= Altitudes

F= Frequencies

T= Transponder Setting

These are just some examples, but there are many more. You can do an online search.

Decision making: the cognitive process of selecting a course of action from multiple alternatives.

We all make decisions every day; the extent to which they will be compromised depends on the environment, such as stress, fatigue, weather, colleagues, etc. In aviation, you have to think ahead; the aircraft cannot be stopped; each phase of the flight needs to be ready before the other begins. The approach phase must be briefed already before starting it. By anticipating the airplane's movements, you'll be better equipped and more secure to respond to any sudden changes.

You are one decision away to make a significant impact on safety.

✈ Situational awareness

Situational awareness: is the perception of the elements in the environment within a volume of time and space, the comprehension of their meaning and the projection of their status into the near future.

A few cognitive factors come across maintaining an excellent situational awareness, such as perception, understanding the data the instruments are providing, how all the information is processed together, etc. Other factors, like experience and knowledge, will also help improve situational awareness.

Concentrating too much on a single task might cause you to lose situational awareness. For instance, you might be following a checklist for a technical failure. You might suddenly find yourself unsure of where the aircraft is heading, is the airplane climbing or descending? How is the fuel situation? Where is the terrain located regarding my position?

The most common loss of situational awareness accidents is always CFIT (control flight into terrain). In December 1995, American Airlines flight 965 crashed in Colombia after the crew had mistakenly placed the aircraft into a collision course with a 3000-meter mountain after losing awareness of the surrounding terrain.

✈ Workload management

Workload management is the ability to perform and delegate a task in a strategic priority or order for a safe and optimal outcome.

As the workload increases, the safety margin decreases. In aviation, the following is a rule that every pilot must always remember.

> Aviate, Navigate and Communicate

This implies that flying should always take precedence over other activities, like communicating. Crew Resource Management (CRM) is crucial for managing workloads effectively. In most commercial aircraft, with two pilots on board, tasks can be delegated to distribute the workload efficiently.

Plan, prioritize, and schedule tasks, manage time effectively, review and cross-check.

Inexperienced pilots often struggle with managing high-pressure scenarios. It's apparent when a novice pilot is getting overwhelmed and beginning to mentally shut down.

This situation is risky since they can't offer much assistance. At that point, you're essentially on your own. It's always best to approach things one step at a time.

On a side note, automation is a great tool to help you. Use it as much as you can, depending on the circumstances, to reduce your workload. Also, use ATC, ask them whatever you need help with the weather, runway in use, vectors, or any other relevant information required.

✈ Multitasking

Multitasking: is the simultaneous execution of two or more tasks at the same time.

We must differentiate two types of multitasking:

- Attention switching: We cannot look at two instruments simultaneously or listen to two conversations at the same time. We need to switch attention constantly while flying, as well as when reading the checklist, where you switch attention from the checklist item to the relevant switch in the flight deck, psychologist refers these as bottlenecks. Basically, you cannot think of two different things at the same time.
- Simultaneous performance: this type of multitasking becomes possible when no bottlenecks are present,

performing two tasks simultaneously, such as flying the aircraft and communicating to ATC on the radio. With practice, this becomes automatic, like driving a car while listening to music at the same time. The resources that require the pilot's attention are free to welcome thinking and talking, for example.

Distractions, interruptions, and fixation in one task can become a safety issue when multitasking.

✈ Confidence

Confidence: A feeling of trust or belief in your abilities and skills to perform a task

Imagine being a passenger on a flight knowing the pilot is not confident enough in their skills and knowledge to fly you safely from point A to B. How scary would that be? Airlines are looking for confident pilots, but of course, not every pilot was confident initially. Confidence builds up through training and experience. A combination of knowledge and developed skills will make a pilot confident.

It's similar to taking a test without sufficient preparation. The more you study and practice, the more assured you'll become. It's beneficial to reflect on areas for improvement, providing a foundation for building confidence. This is a common practice among pilots, not just during training but also after a typical workday as we drive home. We reflect on our performance, especially focusing on how well we executed our approach and landing.

✈ Adaptability

Adaptability is the ability to adjust to changing circumstances quickly.

In the day-to-day aviation world, there are so many variables requiring pilots to be flexible and adaptable to circumstances. Some days the plan is to return home by midnight but instead, you find yourself spending the night in another country Numerous factors like airport closure, delays, weather, sick passenger, can disrupt your plans. These delays sometimes affect hundreds and even thousands of passengers costing the company a significant economic loss, so it is expected from you to devise a solution that does not compromise safety and minimizes the delay impact.

Now, how do all these interpersonal skills come together when conducting a flying check or on a simulator assessment? The instructor is looking to evaluate how you interact and collaborate with your colleague while handling situations or emergencies, using the appropriate procedures and knowledge. How you will prioritize tasks (flying, navigate, communicate in this order) or delegate them at the same time, while maintaining your awareness of your location, your surroundings, and the current status of the aircraft.

If you can manage all this in one check, then you're definitely ready to fly. The airline will entrust you with the lives of hundreds of passengers and the responsibility of operating an expensive, sophisticated machine.

Hazardous attitudes

On the contrary, some hazardous attitudes and personalities can be very dangerous in a cockpit environment, and its best to "leave" outside the flight deck.

The following are the five most recognized:

NAME	DESCRIPTION	ANTIDOTE
Anti-authority (ego)	"Don't tell me..."	Follow the rules; they're usually right.
Impulsivity	"Do something quickly!"	Not so fast- Think first!
Invulnerability	"It won't happen to me...."	It could happen to me!
Macho	"I can do it."	Taking chances is foolish
Resignation	"What's the use?"	I'm not helpless

Source: www.aopa.org

A study examining FAA-defined Hazardous Attitudes in U.S. air carrier flight crew related accidents between 1991-2018 found that anti-authority and invulnerability were present in 92% and 68% of aviation accidents, respectively.

One example of such hazardous attitudes is Garuda Indonesia flight 200 on March 7th, 2007, where the crew overran the runway, crashing into a rice field after the Captain got fixated with the landing. Despite the excess of speed and steep descent, he ignored repetitively the signs to abort the landing and go around by the Fist Officer, as well as by the airplane.

Keep in mind that these dangerous attitudes have no place in a professional cockpit, and failing to address them can result in loss of life.

8 – FINANCING TRAINING

Financing is probably the biggest obstacle to become a pilot. Not everyone can and is willing to pay such a significant amount, especially without a job guarantee. There are risks associated, but it is possible. There are a few different routes to it. In this chapter, you will find a few options available to finance your dream of taking the skies!

The total pilot training cost can vary between **65.000 to 140.000** euros or dollars approximately. This number does not include the type rating (commercial aircraft type). Don't throw in the towel just yet! We'll deal with the type rating later, as many airlines offer excellent financing options for it.

Certain schools offer packages that cover accommodation, three meals a day, uniforms, iPads, books, headsets, and more. It's important to consider these inclusions because if they're not provided, you'll have to purchase them separately and account for additional expenses.

The different alternatives and options to find financing for pilot training are:

- ✈ **OPTION A:** family finances pilot training. Best and most straightforward option!

✈ **OPTION B:** work somewhere for 3 to 5 years or as long as it takes, and save, save, save every cent towards flight training.

✈ **OPTION C:** you are able to attend flight school while working because there's a flight school close to your home, and your job offers flexibility to let you attend official exams.

✈ **OPTION D:** a loan. Private student loan, career training loan, federal student loan, credit unions loan, etc. Ask for a student loan from the bank, start returning the money once you get your first job. Most banks will require some kind of avail. *Some credit unions, like APCU/Center Parc, offer specialized flight training loan programs with flexible terms.*

✈ **OPTION E:** get yourself into an airline "ab-initio" program that finances the training, the only option that guarantees a job (provided that you successfully pass the training).

✈ **OPTION F:** state finances training. You need to check your own country for a state program. Some countries like Poland have this option. Another alternative is military pilot training.

✈ **OPTION G:** personal savings.

✈ **OPTION H:** through the chosen flight school. Most flight schools have agreements with pilot unions to help you finance your training.

For me, it started as option B, continued with option A. To finance the type rating, I used option D. You can even use a combination of all the options like I did.

If you're short on funds, a practical choice is modular training, as it lets you work while progressing through the modules (licenses) your own pace. If you find yourself low on funds, you can work for a while and save up for your next license. I've come across incredible stories of people whose parents sold a second home in the mountains or received their wedding money early to pay for flight school (sounds unbelievable, doesn't it?) or even had their grandmother give them a substantial sum of money, and so on. A colleague of mine had her husband pay for her the training! When there is a will, there is a way! If this is truly your desire, you'll either discover a path or put in the hard work, even if it requires juggling two jobs and giving up weekends for a couple of years.

Before you begin the full training, you must devise a cost-effective plan, as payments are required in advance. If you opt for an integrated course (excluding FAA), you'll need the full payment upfront, which is typically about 20% more costly than the modular option. If you opt for modular training, make sure you have enough funds to secure your PPL first. Should things become too challenging after obtaining your PPL, remember there's a silver lining: you now hold a license that permits you to fly, even if it is just for recreational purposes.

Why is flying so expensive? Because you need to build hours on a small plane. That's where the big part of the budget goes. For the school to have these beautiful light airplanes implies paying airport fees, regular maintenance, replacement of aircraft parts, fuel, and more. You will spend around 150 to 450 euros or dollars an hour of flight training depending on whether it is a single-engine or multi-engine airplane, with or without an instructor. Simulators are also pricy and costly to maintain since they need to be housed in a non-humid environment, with air conditioning and de-humidifiers constantly running. All panels and instruments are electronic and must be kept on a temperature and humidity controlled ambient.

Remember to account for additional training expenses that will increase your budget. These include uniforms, medical certificates, medical insurance if you choose to train overseas, landing and exam fees, and equipment like logbooks and a flight bag with all necessary accessories.

Financing flight training places a significant responsibility on the student pilot to manage a substantial sum of money. I've witnessed individuals misuse all of their parents' money without accomplishing anything. If you're a parent reading this guide for your child, I highly suggest overseeing their school funding.

Type rating:

A type rating (TR) allows a pilot to operate a particular model of commercial aircraft. Therefore, a pilot with a TR for a B737 cannot fly an A320 unless they also have a TR for that specific aircraft.

There's no need to be concerned until you've successfully cleared an airline interview. Typically, airlines will cover the type rating costs, secured by a bond of three to five years, depending on the aircraft type. If you choose to leave before the bond period ends, you'll need to repay the amount specified in your contract.

For instance, if you decide to leave the airline during your first year of employment, you'll need to repay 20,000. If you leave in the second year, the amount is 15,000, and in the third year, it's 10,000, decreasing progressively. Some airlines will subtract this from your salary over the first 3 to 5 years. The cost of type ratings in Europe varies between 20,000 and 65,000 euros. In the United States, obtaining a type rating costs anywhere from $12,000 to $42,000. This variation depends on factors like the aircraft's size, power, and system complexity. For instance, a type rating for a Boeing 737 is approximately 20,000 euros in Europe and 13,000 dollars in the U.S.

The TR course Is split into two sections: theoretical knowledge and practical application. The theoretical segment covers aircraft limitations, systems, and operations. The practical portion takes place in a simulator, culminating in a skill test for European pilots or a check ride for FAA pilots. After completing this, you will proceed to fly the actual aircraft.

🖉 Please read Piotr's personal journey about funding his pilot training while balancing his job as a flight attendant and raising two children at the same time on chapter 19.

9 – CHOOSING A SCHOOL

With financing options in hand, it's time to select the ideal school! Choosing the right school can greatly influence your training and future as a pilot. Some students fail to complete their training because the school went bankrupt, they had a bad experience with instructor or, the aircrafts maintenance was poorly. Below are the steps to help you make the most appropriate decision.

Things to consider when looking for a flight school:

✈ **Location**:

Make sure the school either has a campus (if you plan to be a full-time student) or is located in a place where you can easily find accommodation and transportation nearby.

Weather plays a crucial role. Studying in southern Europe is often more practical than in the north (unless the latter suits you better), as you can maximize your flying time without having to cancel flights due to adverse conditions like rain, low visibility, snow, or strong winds. Some regions offer year-round flying opportunities, while others are limited to a few months for training.

✈ **Fleet**:

A school of substantial size will typically have at least ten aircraft for students to use. It's best to avoid places with only three or four planes, especially when one might be undergoing maintenance, and the school has many students. In such situations, it would be challenging to rent an aircraft for three or four hours for a cross-country flight, as they need to be available for other students' lessons.

✈ **Maintenance**:

Some schools have their own maintenance, which is beneficial because if a plane experiences technical issues, it could lead to financial losses and delay student training. This is especially problematic if the aircraft needs to be sent elsewhere for repairs.

✈ **Reputation**:

Ensure the school you chose has been in the business long enough, and that some of its former students are already employed by airlines.

Having a specific school's name on your resume can expedite your job search. Airlines appreciate certain flight schools for the way they train and educate their students, especially if they emphasize uniformity, airline discipline, and adherence to standards.

✈ **Facilities**:

The school ought to have at least one simulator and should supply the course materials when you enroll or, at least have a store where you can buy them. It should also feature large classrooms with projectors, a flight dispatch area for checking the weather, and route planning with charts, among other facilities. If you're spending that much money, you should expect at least a decent school with excellent facilities to help you succeed!

The following step Involves researching online schools In the region where you wish to receive your training—this could be in your hometown, somewhere in southern Europe, or in the United States.

To simplify the school search, follow the next steps:

1- Decide where you want to study
2- Conduct online research about all the school programs in that area, including prices and reputation. Check forums and possibly ask other enrolled students about the school.
3- Narrow your search to the top 3
4- Go visit them. Schools will offer you a tour of their facilities and guide you around (it's a great opportunity to chat with other students).
5- Pick one. Begin the enrollment process with the chosen school and program.

Ensure that you thoroughly read and comprehend the contract before signing it. Don't hesitate to ask any questions, even if they seem silly. If you're a teenager, it's wise to bring your parents with you for guidance

<u>Questions you should be asking when you visit the flight academy:</u>

- ✈ What percentage of graduate students are currently working for an airline? At my U.S. school, the entry hall was filled with hundreds of photos from students who are now professional pilots, taken in the cockpit. To me, that was clear evidence of the school's success.
- ✈ Fleet size and if the school carries its own aircraft maintenance
- ✈ Cost of repeating an exam and exam fees
- ✈ Price for additional flight hours with an instructor and whether they offer hour package
- ✈ Average ATPL or ATP grade in the school
- ✈ Are you in contact with any airline?
- ✈ How are the payments made? Pay as you go or school account with a minimum amount. If that is the case, make sure you transfer small amounts around 10k to 15k. Avoid depositing large amounts or paying for any course in full if it involves a substantial sum.
- ✈ Does the school have a financing program?
- ✈ What simulators do you have?
- ✈ Make sure they provide a tour of the school facilities and campus if you're thinking about staying in the school accommodation.
- ✈ Is there a cafeteria or dining area at the school where I can eat between classes?
- ✈ Additional expenses while studying in Europe, what are the airport landing fees, and so on?

Future Pilots Checklist

STEPS TO CHOOSE THE RIGHT SCHOOL	
1	Pick up a location ☐
2	Do extensive online research based on the chosen location ☐
3	Narrow your search to the top 3 ☐
4	Go visit them ☐
5	Choose one and enroll! ☐

✎ **NOTE** For EASA future students, take a look at www.theairlinepilotclub.com for partnered flight schools.

10 – OTHER PILOT SCHEMES

University degree with an ATO

Certain flight schools provide an option to finish your training with a degree in aviation management. This will give you a broader understanding of the entire process, including airline flight operations, coordination of ground and aircrew, airline marketing, and the management of airline and flight planning.

This degree opens doors to roles such as air operations chief, quality manager, head of crew training, and ground and air operations manager. These are just a few examples of the career opportunities available with an aviation degree.

Is it really necessary? Airlines don't mandate a degree in aviation for someone to work as a pilot. Sure, having additional knowledge will do no harm. However, you may consider it as a backup plan or an alternative option, in case you decide to pursue a different direction in aviation or if you lose your medical certificate.

These degrees come with extra expenses on top of the regular pilot training fees. By spending approximately 2500 to 5000 euros/ dollars, you can enhance your pilot license with an academic degree.

Airline cadet scheme

Certain airlines offer cadet programs designed to recruit individuals who lack a pilot license or flying experience. After a thorough selection process, these candidates are trained to become first officers. The airline will direct you to a partnered flight school. Upon graduation, you are assured a job, as long as you complete the training successfully and exhibit a positive attitude and good conduct (since no company desires problematic or irresponsible employees).

Certain companies might deduct the training costs from your salary on a monthly basis once you begin flying. Others might ask for payment upfront for the training, but by then, you have already cleared their selection process and are "assured" of the job upon completing the training.

I will refer to the MPL license* (which is not governed by the FAA); airlines collaborate with flight schools to train their cadets.

*For a complete description of MPL, see chapter 13.1.2.

The usual process when going for a cadet program or sponsorship is:

     1. Check eligibility requirements

     2. Submit online application

     3. Initial screening

     4. In-person assessments

     5. Final selection

Government-sponsored programs

Certain nations offer the opportunity for individuals to train as pilots. The selection process is highly competitive, but it's an excellent opportunity! Be sure to verify if your country provides this program and if you meet the necessary qualifications. If you do, begin your preparation diligently, as there will be significant competition!

Military pilot

Did you know, in the past, almost all airline pilots were recruited from the military?

Today there are still some military professionals that join the airline industry after many years of service. Remember, the lifestyle and job in the airline sector differ greatly from military life. It's important to consider if this new path aligns with your preferences. The military involves more risks, such as flying into hazardous and conflict zones, managing threats, and handling weapons.

If you want to pursue that path to reach the skies, be aware that the government funds the training. However, you'll need to successfully navigate a challenging selection process to get in. It's also a fantastic chance to become a pilot and fly impressive machines!

Pay-to-fly scheme (not under the FAA)

The 'Pay to Fly' (P2P) scheme is an unconventional employment practice where pilots pay to work rather than earning a salary. Though not common, this arrangement enables inexperienced pilots to gain experience with a company and obtain a type rating. Typically, pilots who participate in 'Pay to Fly' do so because they struggle to secure a job due to unsuccessful interviews, poor training records (such as multiple ATPL failures), or other similar reasons.

It Involves covering the cost for a type rating that Iles a minimum of 250 or 500 hours on a specific airline model. However, there's no assurance that the airline will hire you once you've completed these hours. This is primarily done on B737 and A320 aircraft. The fees range from 30,000 to 50,000 euros, which are in addition to your license expenses.

Pay-to-fly is a controversial practice that faces significant criticism. While it is legal in most countries, France considers it exploitative for non-employees to pay an airline to gain work experience. Each day, more organizations and pilot unions oppose this practice. The European Cockpit Association (ECA) firmly rejects these practices It seems that anyone with enough money can become a pilot. Companies endorsing this practice are willing to allow anyone who can pay to sit in the cockpit and fly passengers. They don't thoroughly examine your skills, abilities, or training background, as they have turned this practice into a business.

I suggest you "earn" your wings through other means, such as working as a flight school instructor, instead of pursuing this path.

It's important to note that cadet programs often open and close their application windows throughout the year. Aspiring pilots should regularly check airline career pages and flight school websites for the most up-to-date information on program availability and application deadlines

> ✎ **NOTE:** Do an AI or Google search with the following key words: "airline pilot cadete programe Europe" Change Europe for USA, for example.

11 – MEDICAL CLASS 1 CERTIFICATE

Having understood the qualifications and abilities airlines seek in a professional pilot, your next step is to obtain a medical certification. This is essential before you can begin training or apply for a job. Detailed information about the medical examination can be found in the EASA and FAA chapters, along with the procedures and requirements.

11.1 Why is it needed?

Picture a pilot suffering from schizophrenia or frequently losing consciousness while operating a $100 million aircraft filled with passengers at 38,000 feet. No airline would willingly take such unnecessary risks when it comes to the safety of its customers' lives.

When jets began to be pressurized, the danger of rapid decompression and the onset of hypoxia became a significant area of research. Although initial studies commenced during World War I, it was not until after the war ended that it gained more prominence as a subject of study. By the time World War II occurred, the first pressurized military airplanes had been developed, enabling them to fly at higher altitudes. Human bodies are not built to operate effectively in such hostile conditions, characterized by minimal humidity, radiation, vibrations, and the presence of acceleration forces (G forces).

In 1928, the Aero Medical Association of the United States was established, followed by the creation fo the United States Air Force Medical in 1949. In Europe, the United Kingdom was the pioneer in researching human body physiology at high altitude through RAF (Royal Air Force).

According to the book Fundamentals of Aerospace Medicine by Roy Dehart, defines:

"The goal of the Aero Medical Examination is to protect the life and health of pilots and passengers by making a reasonable medical assurance that an individual is fit to fly. These specialized medical exams consist of physical examinations performed by an Aviation Medical Examiner, doctors trained to screen potential aircrew for identifiable medical conditions that could lead to problems while performing airborne duties. "

Where do we stand today? Modern aviation medicine encompasses not only the impact of pressure and high altitude on the body but also extensively examines human factors. Elements like stress and resilience can affect a pilot's performance, thereby influencing aviation safety.

A medical certificate will confirm that the pilot is both mentally and physically capable of performing their duties safely.

11.2 Who does it, and who must have one?

The doctors conducting this test are certified aeronautical doctors, abbreviated as AME (Aeronautical Medical Examiner). They adhere to the guidelines set by relevant agencies (such as FAA, CASA, EASA, etc.) to determine what should be tested and to assess whether the candidate is deemed fit or unfit to fly.

Any holder of a PPL, CPL, ATP(L), or FI license, along with flight engineers, flight navigators, air traffic controllers, and flight attendants, is required to have a valid medical certificate issued by an AME. The classes of these certificates (Class 1, 2, or 3) vary depending on the duties involved and are regulated by the FAA or EASA. These regulations are detailed in EASA 14.1.1 and FAA 14.2.1.

11.3 What tests can I expect?

You can expect on your initial examination a more thorough and complete check. After obtaining your class 1, you'll need to renew it annually. The examination will consist of a variety of checks such as:

- ✈ hearing
- ✈ height and weight measurements
- ✈ blood test for anemia and cholesterol
- ✈ general physical examination
- ✈ urine and use of drugs

- ✈ history of previous accidents or illnesses
- ✈ electrocardiogram ECG
- ✈ lung function
- ✈ psychological assessment
- ✈ vision and color-blindness

Why is this final test crucial? It is vital for a pilot because it distinguishes green from red. These two colors convey distinct information on the instruments, like recognizing terrain or executing a TCAS RA escape maneuver.

Another key reason is to recognize the aircraft at night by its navigation lights, with the red light on the left and the green light on the right.

To obtain an FAA Night Rating, you must accurately identify the first 9 out of 15 plates. You can "test" yourself by searching for the plate numbers online using Google.

Look for an example in the figure below. (Look it up on Google in full color as black and white will not work)

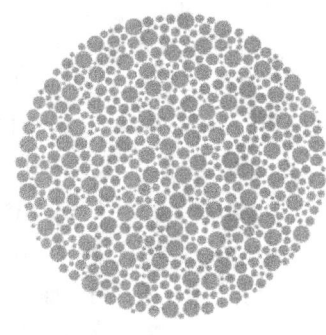

The AME will also inquire if you are using any psychoactive substances, like medication for anxiety, depression, sleep aids, or muscle relaxants, for instance.

Once the medical examination is done, the doctor might issue the medical certificate on the same day, but if further tests are required by a specialist, you will need to do them first. It's advisable to schedule your medical exam well before the expiration date to avoid being taken off the flight schedule.

11.4 Failing the test

There are three possible outcomes on your medical examination:

- ✈ **Issued**: you might have a restriction or an unrestricted medical certificate
- ✈ **Denied**: this is not necessarily a permanent condition always
- ✈ **Deferral**: further testing is needed

Certain conditions may prevent you from passing your medical examination, including color blindness, epilepsy, head injuries, large obstructive kidney stones (which many pilots seem prone to developing), and arrhythmias, among others. However, not passing the initial class 1 (1st class) certification doesn't mean it's over. Don't worry if you have diabetes or another medical condition. If you're uncertain, consult the AME, who can inform you whether your health meets the standard criteria, and if it doesn't, what steps you can take to improve your health status.

At times, all you need are a few expert evaluations or to undergo treatment for a few months. Whatever the case, it doesn't mean it's over!

11.5 Time, cost, and validity

If you're undergoing an initial check-up, you'll likely spend the entire day at the clinic or split the procedure over two days. In contrast, the annual renewal exam lasts around four hours. It's essential to arrive early in the morning with an empty stomach and a full bladder. Blood and urine samples will be collected first, followed by the remaining tests.

The medical examination cost is covered in EASA and FAA chapters, as well as the validity.

11.6 Limitations

EASA restrictions are detailed in chapter 13.1.1 of the EASA medical class 1, while FAA restrictions can be found in chapter 13.2.1.

It's wise to maintain a regular exercise routine. If you're not fond of the gym or working out, aim for at least an hour of walking each day. Regular visits to a chiropractor for adjustments are also beneficial. Prolonged sitting, often accompanied by muscle tension, can significantly strain our backs and spine.

11.7 Tips before the examination

1- I recommend using the same AME on all yearly examinations to establish a relationship. It is expected that your AME will know better than anyone your medical history and will look upon your best interest to help you with any problem.
2- Do not wait until your medical certificate is due. Go a few weeks early just in case your AME requires further testing or a specialist's report. This way, you can maintain your valid certificate during that period without being removed from your company's flight schedule.
3- Bring any medical records you have. Perhaps you were hospitalized for an infection earlier that year, had your appendix removed, or underwent treatment for another medical condition.
4- If you wear glasses, contact lenses, or use hearing aids, make sure to bring them along. You will be tested while using them.
5- Ensure you're ready for your exam by following these tips: Get a good night's sleep, eat a balanced

meal, steer clear of alcohol for your urine and blood tests, and limit caffeine intake since you'll be undergoing an ECG. It's all pretty straightforward!

6- If you have any doubts, contact your AME early.

> 🖉 To learn more about pilot incapacitation, read the blog titled **'Pilot Incapacitation'** on the books website to discover more information and examples.

12 – LANGUAGE BARRIER IN AVIATION SAFETY

12.1 The implementation of standard English in aviation

In 2008, the International Civil Aviation Organization (ICAO) established a minimum English Language Proficiency (ELP) requirement of level 4 or higher for all pilots flying internationally and air traffic controllers working at international airports handling international flights, with a compliance deadline set for 2011. At the time this English proficiency standard was set, only 17 out of the 190 countries surveyed by ICAO had responded.

This was due to the lack of a standardized aviation English for all international pilots, which made flying unsafe and contributed to several incidents in the past. For instance, in June 2007, Flight LOT 282 wandered the skies over London, unsure of its position after an instrument failure. The crew struggled to comprehend the fundamental instructions given by the controller and failed to effectively communicate the nature and extent of their problem. A 2008 study reported by The Times revealed that only 15 out of 800 Polish pilots successfully passed the ICAO English Language Proficiency test.

Even today, standard aviation English continues to cause incidents. In 2017, the UK CAA identified 267 incidents resulting from miscommunication in UK airspace over the prior 28 months. Although pilots undergo all their training in English and are required to have an ELP certificate, their English proficiency is also evaluated during interviews and each simulator assessment.

However, English is not the sole language used in aviation. There are five other official ICAO languages: Arabic, Chinese, French, Russian, and Spanish. For instance, if you travel to Spain, you have the option to communicate in either in English or Spanish. In China, air traffic controllers communicate with Chinese pilots in Mandarin. However, there's a risk of missing crucial details and losing situational awareness if you hear communications in languages other than English.

NOTE: You can read the full incident report online.

Is that a good idea? This question has sparked much discussion. Some people support using the native language to ensure safe operations, while others believe that using non-English communication could jeopardize and diminish the situational awareness of surrounding traffic.

> The main goal of these ELP assessments is to ensure that pilots from all nationalities and cultures adhere to international English standards, **prioritizing aviation safety.**

12.3 Levels

The ELP test evaluates your ability to carry out radio telephony tasks in English proficiently. Airlines require you to have an ELP certificate. The test has 6 levels, and you need at least level 4 to work in the industry. Achieving level 6, the expert level, means your ELP certification will never expire, equating your proficiency to that of a native English speaker. The administration sets these standards, and air traffic controllers must also obtain an ELP certification.

LEVEL 1	pre-elementary	
LEVEL 2	elementary	
LEVEL 3	pre-operational	
LEVEL 4	operational	(expires after 3 years)
LEVEL 5	extended	(expires after 6 years)
LEVEL 6	expert	(No expiration)

The test evaluates comprehension, fluency, structure, vocabulary, pronunciation, and interaction. If you can maintain a conversation in English, you should not be too worried.

If English is your native language, you have even less reason to be concerned!

12.3 Suggestions

What should you do if you didn't score a level 4 on your test or if you're struggling with aspects like pronunciation? My top suggestion is to immerse yourself by living and working in an English-speaking country, even if it's just for a summer or a few months.

If you decide to do this, be careful not to spend too much time with people from your own country, as it will only prolong the learning process! If that's something you're unwilling or unable to do financially right now, consider joining an English academy for an intensive course. Attending class just a few hours weekly won't speed up your learning. At home, you can watch YouTube videos or your favorite Netflix series and movies in English with English subtitles. It may sound strange, but try to think in English as well.

And last but not least, and an excellent suggestion is to listen to www.liveatc.net. For instance, you can listen to the JFK approach. Pilots must repeat back all instructions from the controller, so try to grasp what's happening in real-time.

During my ATPL studies in Florida, I used liveatc.net to familiarize myself with the accent. It was my first experience communicating with ATC in English. While I was flying, my parents in Spain occasionally tuned in to listen to me.

It's a great tool

SUGGESTIONS

You can repeatedly practice the ELT test online. Several websites provide online testing options. Here are some recommendations:

- ✈ ILPT.net
- ✈ Level6Aviation.com
- ✈ ICAO4U.com
- ✈ Aero-Language.com

13 – TYPES OF WORLD LICENSES

As a pilot, the type of license you obtain depends on the regulatory authority of the country where you undertake your training. If you complete your pilot training in the United States, you will receive an FAA pilot certificate. However, unless you are a U.S. citizen or find a way to establish residency, securing employment in the U.S. will be nearly impossible.

The situation is similar in Europe; if you undergo training there, you will receive an EASA license from the relevant authority of the country where you trained. However, without a European passport, you won't be able to work in Europe, as airlines require this. Certainly, you can complete the training in any location and then transfer the license to the relevant authorities in your country, such as Colombia, South Africa, or South Korea.

At this point, you might wonder why not simply undergo the training in your own country. The reason is that you'll end up with two licenses, and major airlines such as Qatar or Emirates value and acknowledge the importance of having an FAA or EASA license. The standards are high, and there are worldwide 99 recognized. A Dutch pilot can operate in China with EASA, FAA, or CAAC (Chinese) licenses.

The realm of aviation licenses Is vast and Intricate. I'll provide as much detailed Information as possible to give you a clear understanding and help simplify your decision-making process.

The United States FAA and the European EASA are the two most prominent aviation authorities globally. They are acknowledged worldwide and, having their recognition grants access to aviation opportunities everywhere. I provide a thorough explanation of each one in individual chapters and also, offer a comparison to help you easily identify their pros and cons. Additionally, there are also excellent licenses from countries such as India (DGAC), Canada (TCCA), Brazil (ANAC), Australia (CASA), and China (CAAC), among others. These licenses can be converted to FAA or EASA, or any other relevant authority you may work with.

Remember that maintaining two separate authority licenses means you'll also need two medical certificates. For instance, if you keep both your FAA and EASA licenses up to date, you'll need to obtain medical certificates from both the FAA and EASA.

13.1 EASA

NOTE: If you are completely certain you'll be training under the FAA, you can skip ahead to chapter 13.2.

EASA, the European Aviation Safety Agency, was founded in 2002 and is based in Cologne, Germany. It succeeded the Joint Aviation Authorities (JAA) with the primary purpose of unifying regulations and standards across the European skies to ensure safety. The agency develops and implements rules and guidelines for all 32 member states to apply and follow. The rules and guidelines include type rating, aircraft certification, medical certification requirements, and approved ATOs. EASA also has offices located in North America and Asia as well.

The agency Is constantly reviewing data to enhance aviation safety In European airspace and approve operators from non-EU countries. A list exists of airlines prohibited from entering Europe or flying over EU airspace because they fail to meet international safety standards. You can Google up that list.

The agency is responsible with making sure that all other pertinent authorities, such as the CAA and IAA, adhere to and implement the standard regulations.

13.1.1 EASA medical class 1

An EASA Aviation Medical Examiner (AME) will conduct your examination to issue a medical certificate. If you plan to study at a European school or transfer your license to EASA, please consider the following details.

CLASSES

→ **Class 1**: the most thorough check for ATPL, CPL, and FI holders.
→ **Class 2**: different standards apply for hearing and vision. PPL holders and flight attendants require one
→ **Class 3**: for air traffic controllers

GENERAL STANDARDS

Here are the EASA standards that your AME will use to determine your fitness. (These guidelines are sourced from the UK CAA authority)

→ **Hearing**

It is tested the ability to hear ring tones at different frequencies (pitch) of 500, 1000, 2000, and 3000 Hertz (Hz) within 2 meters (6 feet) on each ear independently.

According to the UK CAA (Civil Aviation Authorities), the hearing should be tested on the initial examination, then every four years until the age of 40, afterward, every two years.

→ **Vision and color-blindness**

You will be tested on the distant vision from 5 to 6 meters, intermediate at 100cm and near vision at 30 to 50cm on each eye and later, both eyes together. First without correction and with correction (glasses or contact lenses) after.

The vision guidelines for EASA are 6/9 (0.7) for distant visual acuity or better per eye, and 6/6 (1.0) or better for both.

As stated by the UK CAA, bifocal lenses and contact lenses correcting for near vision only are not acceptable.

> **Myopia** (avoids centering objects at a long distance) exceeding 6 dioptres
> **Astigmatism** (blurred vision at any distance) exceeding 2 dioptres
> **Anisometropia** (the two eyes have different refractive power, or commonly known as the "lazy eye") exceeding 2 dioptres

The color perception test is called Ishihara and consists of 24 plates, out of which the first 15 should be identified correctly without error. In case of failing the test, further options are available. If you use contact lenses or spectacles, the AME will test you with them.

To conclude the eye test, you will be checked on visual fields, your pupils and optic fundi (reflexes), and ocular motility (range of movement of the eyes).

Find an example of the Ishihara test in chapter 11.3

✈ Blood test for anemia and cholesterol

The AME will check for abnormal hematology and your blood pressure, which should not exceed 160mmHg systolic and 95mmHG diastolic.

✈ Urinalysis

It is checked the glucose, protein, and for any signs of blood with may indicate renal stones or kidney problems.

✈ History of previous accidents or illnesses

The AME will inquire about your past medical history, asking whether you've ever experienced unconsciousness, neurological issues such as seizures or strokes, as well as problems related to hormones, liver, intestines, sexually transmitted diseases, sleep apnea, allergies, tropical illnesses like malaria, or any head injuries, among other conditions. It's quite an extensive list.

✈ Heart

The heartbeat, abnormal palpitations, heart position, is checked along with an electrocardiogram ECG.

✈ Lung function

The spirometry test will test your peak expiratory flow in l/min.

✈ Psychological assessment

During this evaluation, your mood, appearance, and any atypical behaviors or thoughts will be assessed. You will also be inquired about any past mental health issues and substance addictions. Additionally, the AME will ask if you are currently using any psychoactive substances.

✈ General physical examination

The AME will conduct a comprehensive examination of your organs, assessing the complete range of motion in your limbs and joints to detect any signs of weakness, and also examining your spine. They will evaluate your reflexes and balance within the vestibular system. Additionally, the endocrine system will be tested for any hormonal irregularities, and your vascular system will be assessed by checking your resting pulse.

Essentially, the AME needs to ensure you are mentally and physically capable of holding a commercial pilot license. Following the Germanwings crash in 2015, the psychological component has been given increased attention. Your AME will likely ask questions to identify any potential issues or concerns. If you have any, be completely honest about them. We all have different personal issues at different times of our life, affecting our ability to cope with stress daily.

After obtaining your medical class 1 certification, you must discuss all prescribed medications by your general practitioner, with your AME. For instance, some cough syrups are prohibited because they contain codeine, and certain sleeping pills may be disallowed. If any other medical conditions arise or surgeries are needed, the AME must be notified. Additionally, for women, the initial indication of pregnancy should be reported!

NOTE: You should inform your AME about the location and shape of your tattoos, scars, or birthmarks. This information could help with identification in the event of a fatal accident.

To find out if you qualify with conditions like diabetes, HIV, past eye surgery, or a history of cancer, please visit the Easa website for a pdf document called "easy access rules for medical requirements"

COST

General prices for the Class 1 medical range between 80 to 350 euros, depending on the country and the AME. In Greece, it can cost around 80€, Spain around 200€ and reach up to 450 pounds in the UK.

The initial examination typically costs around 700 euros.

DURATION

The certificate remains valid for 12 months. After turning 40, you must renew it every six months when flying single-engine airplanes, and this requirement also applies from the age of 60.

LIMITATIONS

If you use glasses or contact lenses, it's essential to carry an extra pair in your flight bag. If you're considering Lasik eye surgery, it's wise to discuss it with your Aviation Medical Examiner (AME). Personally, I didn't experience any issues since I underwent Lasik before becoming a pilot. However, some individuals have had problems because the surgery can worsen night vision.

There are about 18 limitations to your class 1 medical:

> **TML** Time limitation
> **APL** Valid only with an approved prosthesis
> **VDL** Wear corrective lenses and carry a spare set of spectacles
> **AHL** Valid only with approved hand controls
> **VML** Wear multi-focal spectacles and carry a spare set of spectacles
> **OML** Operational multi-pilot limitation
> **VNL** Have available corrective spectacles and carry a spare set of spectacles
> **OCL** Valid only as a qualified co-pilot. Extension to OML
> **CCL** Wear contact lenses that correct for defective distant vision
> **OPL** Operational passenger limitation
> **RXO** Specialist ophthalmological examination
> **ORL** Operational pilot restriction limitation
> **SIC** Specific regular medical examination(s) contact the medical assessor of the licensing authority
> **OAL** Restricted to demonstrated aircraft type
> **HAL** Wear hearing aid(s)
> **SSL** Special restriction as specified

Some of my colleagues who have a history of cancer and are currently flying without issues. You might receive an OML (operational multi-pilot limitation) restriction, which would prevent you from flying with another pilot who also has an OML or is over 60 years old; you wouldn't be allowed to fly solo either. I have this kind of restriction myself because of some small kidney stones that can pass, and it will stay in place until they're gone. Hopefully, one day!

Keep in mind that certain airlines require pilots to possess an unrestricted class 1 medical certificate, as they operate on trans-oceanic routes. If a medical issue arises, reaching a hospital in the middle of the Atlantic would be impossible.

13.1.2 Licenses

Let's discuss the different pilot licenses available, their requirements, and the privileges they offer, to give you a better understanding of the school modules and course types.

You can obtain your pilot's license in any European member state, regardless of your nationality. The type of license you need varies based on the aircraft you wish to operate or the pilot position you aim to secure. Below is a list detailing all pilot licenses and ratings, including the requirements for each and the opportunities they provide.

- ✈ Airplane
- ✈ Balloons
- ✈ Helicopter (H)
- ✈ Seaplanes (S)

Certain licenses are specific to different categories, such as balloons or helicopters. For instance, you can obtain a CPL or ATPL(h) for helicopters, or a PPL(b) for balloons.

> **NOTE**: The European Aviation Safety Agency (EASA) has established a separate regulatory framework for drones or unmanned aircraft systems (UAS) that differs from the traditional pilot licenses for manned aircraft like airballoons, airplanes, helicopters, and seaplanes.

1) LICENSES

a) LAPL

b) PPL

c) CPL

d) ATPL

e) MPL

2) RATINGS

a) night rating (NR)

b) multi-engine instrument rating (MEIR)

c) flight instructor

d) examiner rating

3) ADDITIONAL TRAINING

a) MCC & JOC

b) MCC APS

Future Pilots Checklist

✈ **LICENSES**

a) LAPL(A) – Light aircraft pilot license

It is a recreational license, allowing you to fly <u>only</u> light aircraft with basic instrumentation and tour motor gliders.

The ground exam consists of 9 multiple-choice tests on subjects such as air law, navigation, meteorology, operational procedures, human performance limitations, principles of flight, aircraft general knowledge, communications, and flight planning and performance.

For the practical part, you must complete at least 30 hours, which includes 15 hours of dual instruction and 6 hours of solo flying, with 3 of those hours dedicated to cross-country navigation covering a minimum of 80 nautical miles.

It has a validity of 24 months. This license is only recognized in Europe. You will be able to carry up to a maximum of 3 passengers and fly airplanes of no more than 2.000kg.

If you want to add an instrument rating, you can upgrade to a PPL and include a night or aerobatic rating.

The criteria for obtaining this license are:

1- Be at least 17 years old (you can begin logging hours at age 14)
2- Obtain at least a 2nd class EASA medical certificate
3- Receive ground training from an authorized ATO
4- Pass a knowledge test consisting of 9 multiple-choice exams

5- Accumulate the appropriate flying experience with a total of 30 hours
6- Be granted a radiotelephony operator's license
7- Pass the skill test successfully with an EASA examiner

b) PPL(A) – Private pilot license

In the PPL course, you'll experience real flight for the first time (assuming you don't have a LAPL). You'll be taught the fundamentals of flying, including how to take off, land, and manage an engine failure. Once your instructor believes you are prepared, you'll fly solo, completely on your own. This is a day you will remember for the rest of your life.

You'll engage in cross-country navigation with a road map, traveling from one town to the next by following visual landmarks such as railways, highways, rivers, and coastlines using visual references. This initial phase of flying is absolutely thrilling!

The ground examination includes nine multiple-choice tests covering air law, navigation, meteorology, operational procedures, human performance limitations, principles of flight, aircraft general knowledge, communications, and flight planning and performance.

For the practical part, you must complete at least 45 hours of flight time. This includes 25 hours of dual flight instruction and 10 hours of solo flying. Of those solo hours, 5 should be spent on a cross-country flight covering 150 nautical miles, with stops at two different aerodromes starting from the departure point.

The training program covers both navigation and fundamental maneuvers such as take-off, landing, turns, and emergency procedures. While the minimum required time is 45 hours, most people complete it in about 60 to 70 hours on average. The aircraft used must not exceed a weight of 5700 kg.

The requirements to achieve this license are similar to the LAPL

c) CPL – commercial pilot licence

It enables you to work in paid commercial air transportation. After obtaining the PPL, most modular students focus on studying ATPL subjects while accumulating flight hours to pass the CPL skill test.

The CPL skill assessment is similar to the PPL test but includes additional exercises and a more limited error margin. Candidates must be at least 18 years old, possess a Class 1 medical certificate, and have logged a minimum of 200 flight hours.

It is valid for 12 months. Theoretical knowledge 1 in the following 13 topics: air law, aircraft general knowledge, airframe, systems and power-plant, instrumentation, mass and balance, performance, flight planning, and monitoring, human performance, meteorology, general navigation, radio navigation, operational procedures, principles of flight and, visual flight rules and communications.

The requirements to achieve this license are:

1. Have at least 18 years of age
2. Obtain at least a 1st class EASA medical certificate
3. Receive and log ground training from an authorized ATO
4. Pass a knowledge test consisting of 13 multiple-choice exams with a score of 75% or more 2
5. Accumulate the appropriate flying experience with a total of 200 hours
6. Be granted a radiotelephony operator's license ELT minimum level 4
7. Pass the skill test successfully acting as PIC with an EASA examiner

NOTE:

1. If you have previously passed the 14 ATPL theory subjects, you do not need to take a CPL theory knowledge test. If you plan on passing the ATPL, please complete the theoretical part previously to the CPL skill test.
2. These conditions are based on EU regulation 1178/2011, dated 3 November 2011, Part-FCL, Subpart D, CPL.

d) ATPL – Air transport pilot license

Future Pilots Checklist

The most advanced type of pilot certification for aircraft.

It allows you to fly in remunerated commercial air transport. You need to be at least 21 years of age and successfully passed all of the following 14 theoretical knowledge exams with a minimum grade of 75%.

The 14 ATPL exams consist of 685 questions on a total time of 20 hours and 45 min of exam:

SUBJECT	# OF QUESTIONS	TIME AVAILABLE
Air Law	44	1 hour
AGK: Airframe, Systems, & Power-plant	80	2 hours
AGK: Instrumentation	60	1:30 hours
Mass & Balance	25	1:15 hours
Performance	45	2 hours
Flight Planning & Monitoring	42	2 hours
Human Performance	48	1.30 hours
Meteorology	84	2 hours
General Navigation	55	2:15 hours
Radio Navigation	66	1: 30 hours
Operational Procedures	42	1:15 hours
Principals of Flight	46	1:30 hours
VFR Communications	24	0.30 hours
IFR Communications	24	0:30 hours
TOTALS:	**685**	**20:45 hours**

Requirements from EASA website pdf "easy access rules for authority requirements for aircrew Part-ARA"

You must pass all exams within 18 months of your first exam sitting. You have a total of six sittings to pass all the exams and no more than four attempts at any one subject within those 6 sittings.

The optimal way to approach the atpls is do all exams in four sittings: four tests in the first three sittings and, on the last sitting, the remaining three. In this manner, you still have two additional sittings to retake any test you didn't pass. Based on information from an AI search engine, the most challenging exams are performance and flight planning, with pass rates of 67% and 65%, respectively.

A pilot who has Iy completed all 14 theoretical exams but, hasn't yet accumulated 1500 hours of flight time will hold a "frozen" ATPL (fATPL). After reaching the 1500-hour threshold and fulfilling the requirements set by the competent authority, they can unfreeze the ATPL license by passing a skill test.

There are additional requirements to unfreeze this license:

1- A total flight time: 1500 hours, including:
   -500 hours in multi-pilot operations on aeroplanes
   -500 hours as PIC under supervision OR 250 hours as PIC OR 250 hours combined (with at least 70 hours as PIC)
   -200 hours of cross-country flight time (at least 100 hours as PIC or PIC under supervision)
   -75 hours of instrument time (maximum 30 hours can be instrument ground time)
   -100 hours of night flying as PIC or co-pilot

2- Pass an ATPL skills test

3- Hold a valid Class 1 medical certificate
4- Demonstrate English language proficiency (minimum ICAO Level 4)

e) MPL – Multi-Crew Pilot License

This license is relatively new (2006), introduced by ICAO to simplify and expedite the training process for a specific airline pilot.

This program serves as an ab-initio and cadet training for a particular airline, designed to mold future first officers according to the company's standards. Typically, the initial stage involves completing an interview and all necessary assessments provided by the airline offering the cadetship. You need to be thoroughly prepared because these tests are highly competitive and challenging. For instance, when 15 first officer positions are available, 300 candidates typically attend the open day. Therefore, ensure you study extensively!

In training, you bypass certain conventional licenses such as the PPL and CPL. <u>The MPL program is designed to take you from 0 hours to a frozen ATPL on an airliner seat as a co-pilot directly in 18 months</u>. The airline discipline is implemented from the beginning of the training, and a significant amount of time is spent in the simulator. This is a very intense training, and the cadet's behavior is carefully monitored as they are already considered employees.

However, you will not be able to fly single-pilot airplanes unless you take the single-engine rating or, until you reach

1500 hours and take the unfrozen ATPL skills test, you will establish a bond with the airline that you are partnering with for this MPL program.

> 🖉 Please read Andrea´s success story doing an MPL program on chapter 19.

2) RATINGS

   a) NR – Night rating

It allows nighttime flying and, it is a requirement for the applicant to have a PPL and CPL license.

It takes a minimum of 5 hours, including 3 hours of dual instruction at least 1 hour of cross-country navigation of 50 km (27nm) with five solo take-offs and landings to a full-stop.

   b) MEIR – multi-engine rating, instruments rating

ME and IR don't have to be pursued simultaneously. You can obtain an IR on a single-engine aircraft, and later, when you complete the CPL, you can earn the MEP rating (CPL ME). Personally, I completed all my instrument rating training on a multi-engine plane (MEIR).

> ***AIRLINE BOND**: A pilot training airline bond is a financial agreement between an airline and a pilot trainee. The airline covers the cost of training, and in return, the pilot commits to work for the airline for a specified period, typically 3-5 years. If the pilot leaves before fulfilling this commitment, they must repay a portion of the training costs.

The multi-engine rating allows you to fly multi-engine piston aircraft. You will practice handling engine failures and restarting the engine while continuing to fly with the other operational engine.

IR rating allows you to fly the aircraft in Instrument Meteorological Conditions (IMC). You will learn and practice using all instruments on board for navigation DME arcs, holding patterns, and shot precision approaches used every day by airliners on big airports such as an Instrument Landing System (ILS) down to a minimum decision height of 200feet.

The requirements are to be in possession of a PPL or CPL license, have at least completed 50 hours, and pass a theoretical knowledge and a skill test.

c) <u>other ratings</u>

TMG: touring motor glider

SEP: single-engine piston land (A) and sea or (S)

SET: single-engine turbo-prop

MEP: multi-engine piston land (A) or sea (S)

TYPE RATINGS: for example, an Airbus320 type rating

d) FI – flight instructor

This rating allows you to teach and instruct in any ATO (Approved Training Organization). It's an excellent opportunity to gain knowledge, accumulate hours, and develop a curriculum that could serve as teaching experience in the future.

You can issue, revalidate, renew licenses, and provide flight instruction on PPL, CPL, IR, NR, even teach future instructors.

You will need to have received at least 10 hours in instruments flight instruction and have completed 20 hours of VFR cross-country. Additionally, hold a CPL.

The theoretical part includes teaching and learning instruction in topics such as the learning and teaching process, training philosophies, applied instruction techniques, student evaluation and testing, lesson planning and preparation, and training administration, amongst many others.

A spin entry, spin, and spin recovery training are also mandatory to obtain the FI rating. The training and the practical test will be from the right seat, the instructor's seat.

e) other instructor and examiner ratings

There are a lot of instructor certifications one can acquire. Flight Instructor (FI) is the most common one, and it is described in detail in the previous section.

The other two most frequently encountered, SFI and TRI, are detailed in this section, along with a few additional ones.

CRI: Class Rating Instructor
IRI: Instrument Rating Instructor
MCCI: Multi-crew cooperation Instructor
STI: Synthetic Training Instructor
MI: Mountain Rating Instructor
FTI: Flight Test Instructor

SFI – Simulator / Synthetic Flight Instructor

A synthetic flight instructor rating enables you to offer MCC training and conduct simulator sessions for the issuance, revalidation, and renewal of type ratings, along with Instrument Ratings for both single-engine and multi-engine airplanes.

An SFI is restricted to Full Flight Simulators (FFS).

No medical certificate is required for this license.

I have streamlined the requirements list for obtaining an SFI certificate:

1- Hold a CPL, MPL, or ATPL on the applicable aircraft category
2- Have completed 1500 hours as a pilot on multi-pilot airplanes
3- Have completed within 12 months three route sectors on the flight deck of the applicable aircraft type, or two line-oriented flight training-based simulator sessions
4- For single-pilot high-performance, complex airplanes, to have completed at least 500 hours and hold or have held an IR rating

*For further information regarding the revalidation and renewal of the SFI certificate, please refer to the EASA website PDF, Part-FCL section 940.

TRI – Type rating Instructor

A type rating instructor certification enables the holder to teach for the renewal or revalidation of an Instrument Rating (IR) and a Type Rating (TR). Additionally, it allows the issuance of a TRI (Type Rating Instructor) or SFI (Synthetic Flight Instructor) certificate, given the instructor has three years of TRI experience and provides MCC training. For multi-pilot aircraft, it also allows for the issuance, revalidation, and renewal of type ratings.

A TRI also can conduct line-training under supervision.

I have simplified the requirements needed to acquire a TRI certificate:

1- Hold a CPL, MPL, or ATPL license on the applicable aircraft category
2- For a TRI (MPA) have completed 1500 hours of flight time as a pilot on multi-pilot airplanes
3- Have completed within the 12 months preceding the date of expiration, 30 route sectors including take-offs and landings as PIC or co-pilot on the applicable airplane type, of which 15 sectors might be completed on a full flight simulator (FFS)

*For further information regarding the revalidation and renewal of the TRI certificate, please refer to the EASA website PDF, Part-FCL section 940.

EXAMINER

An examiner's primary responsibility is to conduct the final skill test and proficiency check. This individual is an appointed pilot and flight instructor who must possess the necessary knowledge, background, and suitable experience to conduct the privileges of an examiner.

They cannot be subject to any sanctions, including the suspension, limitation, or revocation of any of their licenses or ratings during the last three years. *

* For further information regarding the prerequisites for examiners please refer to the EASA website PDF, Part-FCL section 1010.

- **FE**: Flight Examiner
- **TRE**: Type Rating Examiner

- **CRE**: Class Rating Examiner
- **IRE**: Instrument Rating Examiner
- **SFE**: Synthetic Flight Examiner
- **FIE**: Flight Instructor Examiner

ADDITIONAL TRAINING

a) MCC + JOC

These courses are not ratings but additional training, specifically airline oriented. You will be trained in flight deck discipline, the use of checklists, handling both standard and emergency procedures for a complex aircraft, collaborating closely with your partner, and effective communication with one another.

It serves as a smooth transition from a small single-engine piston airplane into a commercial jet flight deck.

They are usually conducted in Airbus 320 or Boeing 737 fix simulators.

The course takes from three to five days, depending on the school.

b) MCC APS

MCC courses are mandatory when applying for an airline job. A relatively very new variation of the MCC course, since it is based on an airline pilot standard (APS). The top

European airlines that hire many cadets have collaborated with major schools to provide an MCC program customized with that specific airline's operational procedures. This can be very helpful if you have a company in mind that you wish to start your career with. The MCC APS will prepare you better for that airline assessment, and you might get preference in the selection process. It will also be easier to transition into the type rating since you will already know the company culture and SOPs.

Unlike the traditional MCC and JOC, this improved version includes training for cadets in interpersonal skills such as communication, workload management, leadership, and teamwork, along with cockpit discipline. This is essential because the student has only flown basic school airplanes that lack complex systems and automation.

After the application of the new MCC APS, the success rate increased by 21%* on cadet interviews and assessments. The process lasts approximately 16 to 24 days, depending on the school.

> ✎ NOTE: MCC courses are mandatory when applying for an airline job.

13.1.3 Modular vs. integrated training

Now you might be considering which path to get started since most schools offer two types of professional training: integrated and modular. Modular training is the traditional route, goes step-by-step, and the pilot grows with each license achieved. On the contrary, an integrated course is a fast-track route to a professional pilot.

If you've decided and want to reach your goal quickly, this program could be suitable for you.

I have compiled a list highlighting the key features of each program to assist you in selecting the one that best suits your needs.

INTEGRATED

- ✈ More expensive, it adds around 20% more than modular
- ✈ Faster, you go from 0 to fATPL in 14 to 18 months
- ✈ Intense due to the short duration, the schedules are tight
- ✈ Once you begin training at a specific school, you must see it through to the end without interruptions, unlike modular programs. You'll likely have the same flight instructor and ground teachers throughout the entire training process.
- ✈ A steep learning curve, gaining all the necessary knowledge about flying while also developing the

skills in such a brief period is demanding! However, it sets you up for the intensity of earning a type rating and understanding airline procedures.
- ✈ Airlines favor this approach because of the steep learning curve and strict discipline established from the start. It allows for easier tracking of students' progress and better monitoring of the training program.
- ✈ Airline main carrier specific. Some airlines have a preference for specific flight schools, meaning that certain national carriers rely on these schools and their integrated programs to train their future pilots.
- ✈ No PPL or CPL skills test required, this means you'll need to pass the single-engine piston exam if you want to pilot a single-engine aircraft or take your parents or friends on a flight. Additionally, you won't be able to fly solo during your freee time.
- ✈ Graduate with fewer hours than modular students; you will finish with about 205 to 210 hours
- ✈ Full time, this course does not allow you to have any free time. You are a student 100% of the time

MODULAR

- ✈ Cheaper
- ✈ Slower, as each license is handled individually
- ✈ Allows you to work while studying, if you are short on funds, this course is perfect
- ✈ Paced training, the learning curve is customized for your needs; if you need to work, you can study

ATPLs through an online distance school, and you have the flexibility to fly less often if necessary. Additionally, if you aren't completely certain about pursuing this path, you can begin with just a PPL license instead of committing financially to the entire training program.
- ✈ Flexibility, perhaps you require additional time to absorb all the new information, or you might need to take a break and return to work before finishing modules.
- ✈ Switch schools if you had a negative experience; there's no need to retake the same rating or module at the same school. Opt for a more professional ATO. I completed my PPL at one school and finished the rest of my training at a different one.
- ✈ Additional hours are necessary for building flight time. To take the CPL test, you need 200 hours of flight experience. Many modular students accumulate flight hours while also preparing for their ATPL exams.

How do you decide on the path to begin your aviation career? Several factors will influence this decision: Are you able to study full-time? What is your financial situation like—can you afford to pay for an integrated course upfront? Additionally, your learning ability is important; some people learn slowly, and for others, English is not their first language.

Your success will rely on how well you can handle the new information and whether your circumstances allow you to manage other tasks or work alongside your training.

MODULAR	INTEGRATED
Cheaper	Expensive
Slower	Faster
Flexible	Intense
Paced training	Full time training
Finish with more total hours	Finish with less total hours
Possibility to change school in between modules	All training done in the same school
Typical pilot training	Steep learning curve
	Airline discipline

DISCLAIMER: In my personal experience, almost all pilots I know have gone through modular training. Back in 2013, when I did my training, integrated courses were not typical. Nowadays are becoming more common and, almost all schools offer an attractive package for integrated training.

13.1.4 International students visa requirements

There isn't definitive information on this subject. The details will vary based on the European country where you pursue your training. The school you choose will offer the most precise guidance on the application process. To the best of my ability, I can tell you that you will need to:

1- Enroll in the school/program chosen
2- Make sure you do have a private medical health insurance
3- Make sure you do have the financial means to pay for the training
4- A clean criminal record from the country you are residing
5- A fitness medical report
6- A valid passport
7- Have the unrestricted right to live and work in the European Economic Area (EEA) or Switzerland

13.1.5 Tips to master your EASA training

- ✈ Make sure you don't fail more than two ATPL exams, as certain airlines review your training records and first-time pass rates to assess your performance under pressure. Some of my classmates were turned down because they failed multiple times.
- ✈ Ensure that you can take the ground exams and ATPLs at the school or in the vicinity before enrolling to avoid any surprises.

✈ Ensure you pass your CPL skill test on the first attempt. Failing isn't the end of the world, but it is an opportunity to demonstrate your capabilities for an important license.

> ✎ **NOTE BRITISH SCHOOLS**: since the Brexit, to undergo pilot training in the UK as a non-British citizen, you'll need to apply for a visa:
> -Obtain a Student visa (formerly Tier 4 visa) if your course lasts more than 6 months.
> -For courses less than 6 months, you may be able to use a Standard Visitor visa.

13.2. FAA

NOTE: If you are completely certain you will be training with EASA, proceed to the next chapter and skip this section.

The Federal Aviation Administration (FAA) was established in 1958, being one of the first agencies in regulating air traffic. With its headquarters in Washington DC, they took over the previous Civil Aviation Authorities (CAA).

The FAA is responsible for regulating civil aviation in the United States and issuing licenses, managing the operation of airports and air traffic. These powers are extended to international neighbor waters by ICAO.

NOTE: The FAA uses the terminology "certified" for a pilot, instead of "licensed". Its meaning remains unchanged. The FAA collects and analyses data for statistics and further studies, like air transportation effects on the environment and noise. The administration also encourages the development of new aviation technology.

Like EASA, the main goal of the FAA is to provide safety to national civil aviation.

The FAA consist of four different organizations:

1. **Airports:** ensures the construction and operation of all airports comply with the federal regulations.
2. **Air Traffic Organization**: it is basically in charge of safely manage air traffic in and out and around the country.
3. **Aviation Safety**: is in charge of issuing all the certifications of all personnel, including pilots, airlines, mechanics, and aircraft.
4. **Commercial Space Transportation:** protects United States assets during the launch and re-entry of commercial space vehicles.

And, the FAA outlines various regulations under different sections of the FAR(Federal Aviation Regulations) manual.

Here are the most remarkable ones:

- **Part 61:** Outlines requirements for getting licenses (also applies to small flight instructing schools)
- **Part 91:** General operation and flight rules like private flying for non-commercial purposes
- **Part 121:** Regional, major, and cargo airlines dealing with commercial air service like scheduled flights with a pre-defined route
- **Part 125:** implies commercial air service of airplanes with the capacity of 20 or more seating places and maximum payload capacity of 6,000 pounds or more
- **Part 135:** on-demand flights and scheduled charter flights such as corporate, government, and helicopter operations
- **Part 141:** Rules for flight schools
- **Part 142:** Rules for training centers

It is essential to point out that military aviation does not operate under any of these regulations.

THE USUAL FAA ROUTE	
1	Private
2	IR
3	Time building
4	CPL + ME
5	ATP
6	CFI
7	Regional airline FO
8	Regional airline CAPT
9	Main line FO
10	Main line CAPT

I would like to comment on how aviation culture in the United States differs from that in Europe and likely from the rest of the world. In the U.S., there are neighborhoods known as "residential airparks" or "fly-in communities," where homeowners have private hangars connected to their houses.

> 🖉 To learn more about Fly-In communities, visit the blog titled **"Runway Living"** on the book's website at www.futurepilotschecklist.com

13.2.1 FAA 1st class medical

CLASSES AND DURATION

In this chapter, we will outline the requirements and FAA guidelines for studying at an American school or transferring your license to the FAA. Please note of the different types of medical certificates:

- ✈ **First-class**: for pilots who exercise airline transport pilot (ATP) privileges.
  - A First-class medical is valid for ATP privileges for 12 months for pilots under 40 years of age. And for 6 months for over 40 years of age.
- ✈ **Second-class**: for pilots who fly for commercial privileges, as well as air traffic controllers and flight engineers.
  - A Second-class medical is valid for commercial privileges for 12 months.
- ✈ **Third-class:** for student pilots, recreational pilots, and private pilots.
  - A Third-class medical is valid for 60 months for pilots under 40 years of age,
  - And 24 months for applicants who are 40 years of age or older.

According to the new regulation, once the 1st class medical privileges expire after a year, they automatically downgrade to a 3rd class medical status. This regulation update does not alter the validity period of 2nd class medical certificates.

STANDARDS

Here is a summary of the FAA's medical standards (last revision in 2006).

Medical Certificate Pilot Type	First-Class Airline Transport Pilot	Second-Class Commercial Pilot	Third-Class Private Pilot
Distant Vision	20/20 or better in each eye separately, with or without correction.		20/40 or better in each eye separately, with or without correction.
Near Vision	20/40 or better in each eye separately (Snellen equivalent), with or without correction, as measured at 16 inches.		
Intermediate Vision	20/40 or better in each eye separately (Snellen equivalent), with or without correction at age 50 and over, as measured at 32 inches.		No requirement.
Color Vision	Ability to perceive those colors necessary for safe performance of airman duties.		
Hearing	Demonstrate hearing of an average conversational voice in a quiet room, using both ears at 6 feet, with the back turned to the examiner or pass one of the audiometric tests below.		
Audiology	Audiometric speech discrimination test: Score at least 70% reception in one ear at an intensity of no greater than 65 dB. Pure tone audiometric test. Unaided, with thresholds no worse than:		

Ear Condition	500 Hz	1,000 Hz	2,000 Hz	3,000 Hz
Better Ear	35 dB	30 dB	30 dB	40 dB
Worst Ear	35 dB	50 dB	50 dB	60 dB

Ent	No ear disease or condition manifested by, or that may reasonably be expected to maintained by, vertigo or a disturbance of speech or equilibrium.
Pulse	Not disqualifying per se. Used to determine cardiac system status and responsiveness.
Blood Pressure	No specified values stated in the standards. The current guideline maximum value is 155/95.
Electrocardiogram (ECG)	At age 35 and annually after age 40 (first-class only) Not routinely required. (second- and third-class only)
Mental	No diagnosis of psychosis, or bipolar disorder, or severe personality disorders.
Substance Dependence and Substance Abuse	A diagnosis or medical history of **substance dependence** is disqualifying unless there is established clinical evidence, satisfactory to the Federal Air Surgeon, of recovery, including sustained total abstinence from the substance(s) for not less than the preceding 2 years. A history of **substance abuse** within the

Medical Certificate Pilot Type	First-Class Airline Transport Pilot	Second-Class Commercial Pilot	Third-Class Private Pilot
	preceding 2 years is disqualifying. Substance includes alcohol and other drugs (i.e., PCP, sedatives and hypnotics, anxiolytics, marijuana, cocaine, opioids, amphetamines, hallucinogens, and other psychoactive drugs or chemicals).		
Disqualifying Conditions	Unless otherwise directed by the FAA, the Examiner must deny or defer if the applicant has a history of: (1) Diabetes mellitus requiring hypoglycemic medication; (2) Angina pectoris; (3) Coronary heart disease that has been treated or, if untreated, that has been symptomatic or clinically significant; (4) Myocardial infarction; (5) Cardiac valve replacement; (6) Permanent cardiac pacemaker; (7) Heart replacement; (8) Psychosis; (9) Bipolar disorder; (10) Personality disorder that is severe enough to have repeatedly manifested itself by overt acts; (11) Substance dependence; (12) Substance abuse; (13) Epilepsy; (14) Disturbance of consciousness and without satisfactory explanation of cause, and (15) Transient loss of control of nervous system function(s) without satisfactory explanation of cause.		

Extract by *www.faa.gov*

When am I not required to have a medical certificate?

As a student pilot pursuing:

- Balloon/glider pilot
- Ground flight or simulator instructor

COST

The price of the medical exam can differ based on the examiner and location, ranging from $75 to $200 for a first-class exam. If a Special Issuance medical exam is required, the cost may increase.

LIMITATIONS

SPECIAL ISSUANCE MEDICAL CERTIFICATION

A special issuance on your medical certificate indicates that you have obtained the certificate, but it includes a "limitation" on the expiration date because of a medical condition. If your AME discovers something during your medical exam that needs additional testing and FAA approval, you'll need to provide additional medical reports so an FAA doctor can make the final decision regarding your certificate. Once approved, the medical certificate will remain valid until its expiration date.

The certificate will be reissued once the necessary tests and reports are submitted, so the quicker you provide them, the less time you may have to wait for FAA approval. If your medical condition gets better, it must be proven through medical tests and reports, and the special issuance or waiver might be revoked. The process can be lengthy and exasperating, yet it's beneficial to have alternatives for flying with specific medical conditions. As reported by Aopa, over 30,000 pilots have obtained certification through special issuance.

The 15 medical conditions listed below are explicitly disqualifying, but most still permit pilots to fly with a special issuance medical certification.

- Coronary heart disease
- Angina
- Myocardial infarction
- Heart replacement

- Cardiac valve replacement
- Permanent cardiac pacemakers
- Diabetes
- Psychosis
- Bipolar disorder
- Severe personality disorder
- Substance dependence or abuse
- Epilepsy
- Disturbance of consciousness
- Transient loss of nervous system function

A special issuance is different from a waiver, known as a Statement of Demonstrated Ability (SODA). The waver is typically issued for pilots with static defects that are not likely to change, such as upper or lower limb amputees or monocular vision (good vision on one eye). The candidate has to demonstrate on a medical flight test that he or she can safely operate the aircraft under their permanent condition. The waiver becomes part of your medical certificate.

Some special issuances might be renewed by the AME when they previously have been authorized by an FAA doctor, such as asthma, cancer, hepatitis C, kidney stones, migraine syndrome, ulcerative colitis, sleep apnea, glaucoma, Chrohn´s disease as well as, some cardiovascular conditions such as bypass, coronary artery disease, mitral & aortic insufficiency, etc. In these particular cases, the airman must have a letter from the FAA which authorizes the AME Special Issuance renewal.

The medical certificate Is valid for only as long as the authorization Is In force.

13.2.2 Certificates

You can get your FAA certificates in the United States, regardless of your citizenship. Different certifications are required based on the aircraft you wish to fly or the specific privileges you intend to obtain with it.

Part 61 or part 141?

PART 61	PART 141
Flexible schedule	Follow an FAA approved syllabus
Customized training program	Structured training curriculum
Minimum of 40 hours to PPL	Minimum of 35 hours to PPL
Minimum of 250 hours to CPL	Minimum of 190 hours to CPL
Often more expensive	Cost-effective for full-time students

The FAA permits flight schools to operate under either part.

Here is a list of all pilot licenses and ratings with the requirements and what you can do with each.

(A) Airplane

(B) Balloons

(H) Helicopter

(S) Seaplanes

Some licenses apply for different categories like balloons or helicopters. You can have a commercial or an ATP(h) on the helicopter category. Or a private(b) on the balloon category.

> ✎ **NOTE:**
> 1- Commercial pilots flying with a valid FAA second class medical in some International Civil Aviation Organization (ICAO) countries may not be considered in compliance with that country's regulations because ICAO does not recognize a 2nd class medical certificate in any commercial operations. (*www.aopa.org*)
>
> 2- Make sure you fill out the FAA online application before you go to the appointment on the following website: *www.medxpress.faa.gov/medxpress/*

1)CERTIFICATES

a) Sport pilot license
b) Recreational pilot license
c) Private
d) Commercial
e) ATP

 f) CFI
 g) additional CFI

2) RATINGS AND ENDORSEMENTS

 a) multi-engine and instrument, seaplane and instructor ratings
 b) endorsements

1) CERTIFICATES

NOTE: If you are convinced about becoming a commercial pilot, skip the sports and recreational pilot licenses and start directly with the private pilot certificate. Unless you're a parent looking to introduce your child to flying either for fun or to see if they want to pursue further training at an early age.

a) Sport Pilot Certificate

The Sport Pilot Certificate is relatively new (2004),

and it was designed for individuals seeking to fly a lightweight, compact, and easy-to-operate single or double-seater aircraft.

Practically, it is the most restrictive certificate since you are limited to fly only light aircraft at low altitudes in your local airport area, never at night time, you can only carry one passenger and airspace class B, C or D.

Future Pilots Checklist

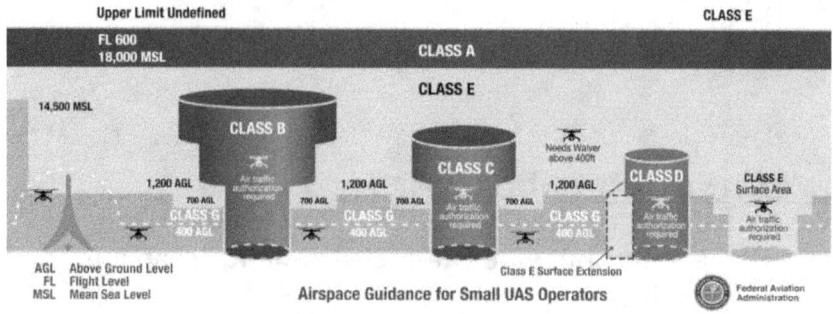

*table extracted from
www.faa.gov/uas/recreational_fliers/where_can_i_fly/airspace_101

For comparisons with Recreational pilot certificate and Private Pilot, there is a table within the next few pages.

b) Recreational Pilot Certificate

The Recreational Pilot Certificate is a bit less restrictive than the Sport Pilot and more than a Private Pilot License. The Recreational Pilot Certificate allows you to fly light aircraft with basic instrumentation and tour motor gliders also. A simple version of the Private Pilot's License (PPL). This certification restricts you to flying only light aircraft at low altitudes within the vicinity of your local airport, and never during nighttime. You are permitted to carry just one passenger and can operate only in airspace classes B, C, or D. *1

It Is the perfect license for those pilots who want to stay outside controlled airports within 50 nautical miles of their home airport and are not interested in pursuing further advanced training afterward.

c) Private Pilot Certificate

The private pilot certificate is the most common type of pilot license. It offers more flexibility than a sport or recreational pilot license. With a private certificate, you can fly at night, access controlled airports, and operate any aircraft in the previously mentioned categories.

The training (Part 61) consists of a minimum of 40 flying hours, including:

- 20 hours of dual instruction, including 3 hours of cross-country flight and 3 hours of night flying, will include 10 take-offs and landings to a full stop and one cross country flight over a total distance of 100nm and, 3 hours of instrument flying
- 10 hours of solo flying including 5 hours cross country flight and one cross country flight*1 of at least 150 nm. With full-stop landings at a minimum of three points
- 3 takeoffs and landings to a full stop at an airport with ATC

> A CROSS COUNTRY FLIGHT is a flight longer than 50 NM

Followed by a knowledge exam of 60 questions, multiple-choice "abc" with a pass mark of 70% or more. The Private Certificate average pass hours is around 70 hours.

For example, most private pilots are certified in the category of Single-Engine Land. You can still add ratings and endorsements like tailwheel, seaplane, complex or high-performance airplane or multi-engine, etc.

This course offers your initial experience with actual flight, provided you don't already hold a sport or recreational pilot license. You'll be taught the fundamentals of flying, how to take off and land, and how to handle an engine failure. Once your instructor deems you prepared, you'll have the chance to fly solo, all on your own.

You will also do cross country navigation using a VFR sectional chart, go from town to town using visual meteorological conditions (VMC) like railroads, highways, rivers, coastline, etc. This first bit of flying is pure excitement!

As well as the sport and recreational Pilot, a Private pilot cannot fly for commercial purposes or economic compensation.

🖉 **NOTE:** Night rating is part of an FAA PPL

	SPORT PILOT	**RECREATIONAL PILOT**	**PRIVATE PILOT**
How many hours do I need to get the license?	20	30	40
What type of aircraft can I fly?	1320 lbs max certificated gross weight, 2 seats	180 horsepower, 2 seat max	Any airplane up to 12500 lbs and maximum 200 hp
Do I need a medical certificate?	No, but U.S. driving license	Yes class 3	Yes
Can I fly at night?	No	No	Yes, with a night rating
Can I fly outside the U.S.?	No	No	Yes
Where can I fly?	Airspace class G or E. Class B, C, and D after special training and endorsement. You are not allowed to fly in class A.	Airspace class G or E. Class B, C, and D after special training and endorsement. You are not allowed to fly in class A.	Any airspace class except class A, unless you have an instrument rating (IR)
Can I fly an airplane with retractable landing gear?	No	No	Yes, with additional training
How many passengers can I fly with?	1 only	1 only	More than 1

Can a charity benefit from a sightseeing flight?	No	No	Yes
Can I fly with less than 3 miles of visibility?	No	No	Yes, in uncontrolled airspace
How long is my certificate valid for?	As long as you have a valid medical	As long as you have a valid medical	As long as you have a valid medical

d) commercial pilot certificate

The commercial pilot certificate allows the certificate holder to receive economic compensation for the flight or transport of passengers or cargo. Applicant must be 18 years of age, have a 2nd Class medical.

The commercial pilot skill test would be like a private pilot test, but with more exercises and stricter error margins. Extended knowledge of professional flight operations is required. A commercial pilot certificate will allow you to fly airplanes with retractable landing gear and controllable pitch propeller, as well as complex systems and high performance (with the appropriate endorsement.)

For the certificate to be valid, your medical needs to be renewed every 12 months. It requires a minimum of 250 hours (part 141, 190h) that includes 20 hours of training and 10 hours solo, including a few cross-country flights: one with more than 50nm from the departure airport, one at least 250nm and, a 300nm total distance with landings at three airports, plus night flying.

I have simplified the requirements list to obtain a commercial pilot certificate:

1. Be at least 18 years of age
2. Obtain at least a 2nd class FAA medical certificate
3. Hold a Private Pilot License
4. Pass a knowledge test with a score of 70% or more
5. Accumulate the appropriate flying experience with a total of 250 hours including solo time part 61 or 190 hours under part 141
6. Pass the oral and practical examination successfully with an FAA designated pilot examiner or inspector up to the FAA Commercial Pilot Practical Test Standards

e) ATP Airline transport pilot

The most advanced pilot certificate. It is mandatory to work for a commercial airline.

You need to be at least 23 years of age, passed a theoretical exam of 100 questions multiple choice with a minimum mark of 70%, and have logged at least 1500 hours, including 500 hours of a cross-country flight, 100 hours of night time, and 75 hours of instrument operations time (real or simulated). Like the EASA ATPL, a restricted ATP with fewer hours is available till reaching the 1500 hours.

Future Pilots Checklist

I have simplified the list of requirements to obtain an unfrozen ATP:

1. Be at least 23 years of age
2. Obtain a 1st class FAA medical certificate
3. Hold an Instrument Rating
4. Pass a knowledge test with a score of 70% or more
5. Accumulate the appropriate flying experience with a total of 1500 hours and be able to demonstrate the experience with a signed logbook
6. Pass the oral and practical examination successfully with an FAA designated pilot examiner up to the FAA Airline Transport Practical Test Standards

This "total time" required is reduced to:

- 750 hours for former military pilots
- 1000 hours for university bachelor's degree graduates
- and 1250 hours for university associate degree programs graduates (from zero hours to commercial)

f) CFI certified flight instructor

Earning a flight instructor's certificate is an excellent opportunity to earn money while flying, rather than paying for it. This qualification enables you to provide flight instruction. You'll not only accumulate flying hours rapidly but also significantly enhance your technical knowledge.

Moreover, it can serve as an excellent stepping stone for transitioning into other instructional roles within an airline.

The theoretical test consists of 100 questions multiple choice on airplanes and aerodynamics, airplane performance, aviation weather, airports, airspaces and ATC, federal aviation regulations, weight and balance, navigation, flight maneuvers, and aero-dynamical factors. The applicant must also pass a theoretical test on the following topics: the learning process, barriers to learning, human behavior and effective communication, teaching methods, planning instructional activity, and critique and evaluation.

A spin entry, spin, and spin recovery training are also mandatory to obtain the CFI. The training and the practical test will be from the right seat, the instructor's seat.

I have simplified the list of requirements to obtain a FI certificate

1. Be at least 18 years of age
2. Hold a commercial pilot certificate or an ATP certificate (with an IR)
3. Obtain at least a 3rd class FAA medical certificate
4. Receive and log ground training from an authorized flight school
5. Pass a knowledge test with a score of 70% or more
6. Accumulate the appropriate flying experience
7. Pass the oral and practical examination successfully with an FAA designated pilot examiner up to the FAA Flight Instructor Practical Test Standards

If a certified flight instructor (CFI) wants to teach Instrument rating (IR), he/she will need to add that rating to his/her license CFII. To teach on a multi-engine, MEI. The same will apply for any other ratings an instructor can teach.

g) ADDITIONAL INSTRUCTOR CERTIFICATES

DPE: designated pilot examiners
PE : pilot examiner
CE : commercial pilot examiner
CIRE: commercial pilot and instrument rating examiner
FIE: flight instructor examiner
PPE: pilot proficiency examiner
SAE: specialty aircraft examiner
SAE: sport pilot examiner
SFIE: sport pilot flight instructor examiner
EAE: experimental aircraft examiner
VAE: vintage aircraft examiner
DME: designated mechanic examiner
DPRE: designated parachute rigger examiner

Source: www.faa.gov/documentLibrary/

2) RATINGS AND ENDORSEMENTS

The correct definition of a rating would be a qualification that authorizes the pilot to operate on that particular aircraft type. Because different aircraft require distinct flying techniques and procedures, obtaining a type rating is necessary. I am certified to operate a Boeing 737 and am permitted to fly only that specific Boeing model. I cannot fly an Airbus 320 unless I obtain a type rating for it.

Consider these more like an add-on or extra to your certificates.

a) multi-engine and instrument, seaplane and instructor ratings

As I mentioned regarding the EASA ME IR, the same rule applies to the FAA—you don't have to complete them simultaneously.

After earning your private pilot license, you can pursue an instrument rating on a single-engine aircraft. Once you finish the commercial license, you can obtain a multi-engine (ME) rating. In my situation, I completed my instrument rating on a multi-engine plane.

If you're training under part 61, you'll need at least 15 hours of dual instruction along with a 50 nautical mile cross-country flight. However, if you opt for training under part 141, you'll only need 35 hours of dual instruction.

During the dual instruction time, you will learn the communication procedures, to do holding patterns, shoot instrument approaches like ILS, NDB, GPS, etc. while flying in IMC and using the instruments as a reference instead of an external reference.

The requirements are to be in possession of a private pilot certificate and, have logged the appropriate flying experience, and have received the ground instruction.

ME allows you to fly multi-engine piston aircraft. You will train to fly with only one remaining engine and how to restart it. You will combine both and learn to shot an instrument approach or enter a holding pattern with one engine inoperative.

- Seaplane: It might be a floatplane or an amphibious aircraft capable of landing on both land and water.
- CFI: Certified Flight Instructor
- CFII: Instruments Rating Instructor
- MEI: Multi-Engine Instructor

b) endorsements

- Balloon: lighter-than-air aircraft without an engine that can sustain lift using a gas buoyancy.
- Airship: lighter-that-air engine driven aircraft. There are very few now.
- Glider: a heavier-than-air aircraft flying without an engine, using the air thermal currents.
- Tailwheel: they have different landing and steering capacities than the typical tricycle gear airplanes
- Complex airplane: an airplane with retractable gear, movable flaps, and a controllable-pitch propeller.
- High performance: aircraft with more than 200 hpw

- ✈ <u>High altitude</u>: aircraft able to fly at higher altitudes and equipped with oxygen systems and decompression procedures.
- ✈ <u>Type rating</u>: specific aircraft type like for example Embraer 190 type rating
- ✈ <u>Spin endorsement:</u> Instructors under training need to have a spin endorsement before taking the final check ride to qualify as CFI.

Ratings are added to your pilot certificate's back, while endorsements are signed stickers by your FAA instructor and placed on your logbook. Find an example of FAA endorsement:

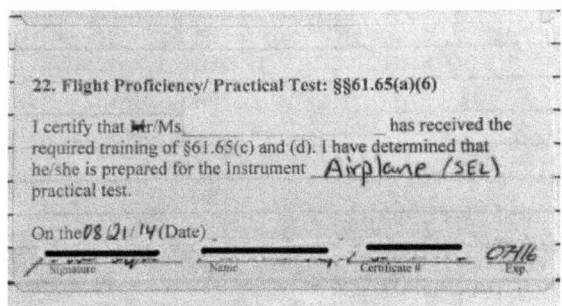

13.2.3 Visa requirements for international students

If you're coming from another country and intend to complete your flight training in the United States, please follow these steps,

which the school will thoroughly explain.

For now, you will be familiar with the procedure and know that it can take from 3 to 6 months to obtain the visa.

Future Pilots Checklist

Knowing this information might help you plan your enrollment in advance.

The procedure isn't complex but, there is a specific sequence of steps to adhere to:

1. Select the flight training school and program you wish to pursue.
2. Complete the paperwork: I-20 form[1] by the school
3. Receive the I-20
4. Pay the fee (approximately 235$) for the I-901 SEVIS[2] and keep with you the receipt
5. Apply for the appropriate visa[3] M1/M2 or F1/F2
6. Book an appointment with the nearest US embassy[4] or consulate (bring with you I-20, passport and copies, financial bank statements, and the I-901 SEVIS receipt)
7. Receive your visa
8. Get TSA clearance[5] (130$ fee) keep TSA clearance copy, fingerprints, etc.
9. Purchase your plane tickets to the United States and begin your training!

Please don't worry! I've outlined the process below and provided detailed explanations for each step.

✈ 1: I-20 form

The I-20 Form is an official document issued by U.S. educational institutions to international students. It:

1. Certifies admission to a full-time study program

2. Proves sufficient financial resources

3. Enables F-1 or M-1 visa applications

It contains key information like SEVIS ID, program dates, and funding details. Students must keep this document throughout their U.S. stay as proof of legal student status.

Visit the website for additional details, where you'll find a helpful Q&A section. If you switch schools within the United States, you'll need to complete a new I-20 form.

✈ 2 **I-901 SEVIS**

This is a fee associated with the Student and Exchange Visitor Information System (SEVIS).

The I-901 SEVIS Fee is a mandatory payment required for most F, M, and J visa applicants before they can obtain a visa or enter the United States as students or exchange visitors. It supports the maintenance of the SEVIS database, which tracks international students and exchange visitors in the U.S

A payment of $350 is required and should be completed on the website only using a credit card. Students must keep the payment receipt as proof, as it will be required during the visa interview and when entering the United States

✈ 3 **<u>types of visas</u>**

B1/B2: it is a temporary, non-immigrant US visa that permits visitors to enter the U.S. for business purposes for up to six months and may be renewed once for up to a further six months. <u>You are not allowed to do pilot training with this visa.</u>

F/M are temporary non-immigrant visas from a consular officer abroad:

M1/M2: it is a non-academic and non-immigrant visa for international students, used for vocational courses only. It allows you to do training for 12 months, which can be extended to another 12 months.

It doesn't permit you to immediately apply for a job after completing the training or to enroll in university studies afterward.

To obtain an M1 visa, you must:

- Be at least 18 years old
- Have an academic education or high school diploma
- Have received an I-20 form from a vocational school in the US
- Have sufficient English knowledge to complete the vocational course in the US
- Have financial proof to fund training and live in the US for 12 months
- Your intention to depart and not remain in the US upon completing the pilot training

This visa allows you to move in and out of the U.S. without restrictions. Rest assured, you can return home for Christmas and then come back to resume your training!

The distinction between M1/F1 and M2/F2 visas lies in that the first one is intended solely for the individual holder, while the later, permits you to accompany family members like your spouse or children.

✈ 4 **embassy appointment**

"A consular officer will interview you to determine whether you are qualified to receive a student visa. You must establish that you meet the requirements under U.S. law to receive a visa." *

- 🖉 Make sure your passport is valid and won't expire within the next six months prior to your visa appointment.
- 🖉 Ensure you have a working email address before initiating any processes and paperwork. <u>Remember to note down the password somewhere safe!</u> To avoid missing deadlines or important paperwork due to forgetfulness. Some people could not find the I-901 SEVIS receipt or had to register twice on the TSA website.

> **NOTE:** F1/F2 is the most prestigious visa offered by less than 10 flight academies in all U.S., usually provided for academic courses and university training.

QUESTIONS YOU WILL BE ASKED:

A series of questions will be posed to you, such as:

- ✈ Why do you want to train in the U.S?
- ✈ What is your academic background?
- ✈ What is your financial situation? Can you afford to pay for training and all expenses for one year?
- ✈ What are your intentions after completing the training?
- ✈ Do you have any family living in the U.S.?

REASONS WHY SOME PEOPLE DON'T PASS THE INTERVIEW

Although I haven't personally encountered anyone who failed the embassy interview, I've heard that some individuals might not succeed because:

- ✈ Their English is not good enough to conduct training
- ✈ Bad attitude during the interview
- ✈ No confidence answering the questions
- ✈ Planning to stay in the U.S. after finishing flight training

- ✈ No financial evidence that shows the candidate can support himself or herself during their time in the US

TIPS TO ASSIST YOU!

Consider these points before you proceed:

- ✈ Show confidence when answering questions; otherwise it might look like you are lying
- ✈ Do not be late!
- ✈ "Try" not to be overly nervous
- ✈ Dress appropriately. It is a business interview, do not show up with your trainers and broken jeans or sports t-shirt; business casual is okay
- ✈ Be polite and never show an arrogant or superior attitude
- ✈ Smile!

- ✈ 5 **tsa clearance**

This procedure began after 911; previously, it was done by private companies. Anyone considering doing flight training in the U.S. has to demonstrate he/she is an American citizen (passport, birth certificate, etc.) or get clearance by the Transport Security Administration (TSA) before training begins.

It is relevant to green card holders, international students, and even tourists.

The process typically takes about 15 days to complete and proceeds as follows:

Future Pilots Checklist

1- Register on the TSA website
2- Wait for the school to receive the request from the TSA
3- Once the school has verified all your documents and has received a deposit6 (**schools will require you to secure a small amount of money before you start; if you get TSA or visa denied, they will refund the amount most likely**), they will confirm the TSA request electronically
4- You will receive an email from the TSA Alien Flight Student Program
5- There is a small fee of 130$ to be paid and a request to submit your digital fingerprints to the school
6- Wait for the final approval to start flight training!

More information on www.tsa.gov

✈ 6 **school deposit**

Before beginning your training program, schools generally require you to deposit a modest sum, typically between $4,000 and $6,000, into your student account to confirm your enrollment for the start date.

13.2.4 Tips to master your FAA training

✈ Ensure you successfully complete the majority, if not all, of your check rides and ground theory exams. This is important because when you apply for a pilot

position or with an airline, they will inquire about any tests you have failed and the reasons behind it.
- ✈ Commit to the full training time if you can, Flying daily is crucial to maintaining your skills and muscle memory.
- ✈ Study ahead of time even before you start the training. Know what to expect. The FAA provides free access to their books online.
- ✈ Certain airlines, particularly major carriers, favor candidates who hold a bachelor's degree, although the degree doesn't necessarily need to be in aviation.

> **NOTE:**
> - ✎ You should complete your TSA application within at least 60 days before the start of your flight training.
> - ✎ Once the final approval has been received, you have 180 days to start training and 365 days to complete it.
> - ✎ Ensure you have <u>health insurance</u> when studying in the U.S.

13.3 EASA vs. FAA

Having undergone training under both EASA and FAA regulations, I can point out some differences, benefits, and drawbacks. I have created a table for each, detailing the minimum required hours, theory exams, medical class type, and more. The purpose of these tables is to enable you to easily compare the EASA and FAA training and address any other questions you may have with just a quick look.

Choosing between EASA and FAA training shouldn't be complicated if you are American or European. If you're a US citizen, it doesn't make sense to obtain a license in Europe, as you wouldn't have the right to use it there without a European passport. If you are neither American or European, then this chapter will help you.

The conversion from FAA to EASA is more complicated due to the extended 14 ATPL theory exams. That's why no American students are studying in European schools. However, if you're European, you might consider training in the U.S. Just keep in mind that you'll still need to pass the 14 ATPL exams to work in Europe and obtain the EASA license.

In my view, training in the U.S. is quicker, more practical, and more dynamic, thanks to the vast number of airports, and it's also more affordable. This is the reason why numerous European students choose to train in the U.S. as I did myself.

You have to choose the best training location based on your budget and where you want or are able to work. Personally, I could have done all my pilot training in Spain, my home country. I have no regrets about completing part of my training in the U.S. I took hour-building flights to New York and New Orleans, had the incredible experience of using airport FBOs, even spending a night in one, and landed at Fort Lauderdale International Airport with the school's small Piper Cherokee (Pa28), taxiing behind a large

American Airlines jet. I significantly improved my aeronautical English and had a ground school instructor who was a Vietnam veteran and a navigator for the Boeing Stratofortress B-52. So many unforgettable experiences.

I must admit that training at home was wonderful. My parents came to the airport to watch me practice touch and goes during my private pilot training. Then, when I passed my commercial pilot license, my whole family gathered at the school to celebrate with a lunch. Flying my mom and having those great experiences was truly special! In any case, whatever choice you make will be excellent. Even after accumulating thousands of hours of flight experience, you can still switch and obtain another license.

I had friends at the school from various places like Colombia, Saudi Arabia, Egypt, Israel, and Mauritius. They completed FAA training and then went back to their home countries to convert their qualifications with their local aviation authorities to fly there.

Please note some of the most notable differences between EASA and FAA pilot training and licenses.:

EASA	FAA
+ complex	+ practical exams
+ difficult theory	all 100% in the English language
the pass mark is 75%	the pass mark is 70%
slower to achieve	faster to achieve
4 question choice "abcd"	3 question choice "abc"
high standards (prestige)	check rides are tougher
+expensive	schools are cheaper

fewer schools and newer	+ airports to land and + approaches
fewer airports where to train	night rating included in ppl
stricter rules	endorsements
only 2 TR at one time	unlimited numb of TR

→ The 14 ATPL subjects are considered the **toughest**, as they come with a question bank of about 15,000 multiple-choice questions, each offering four possible answers (a, b, c, d) and requiring a passing score of 75%. In contrast, the FAA ATP question bank contains roughly 8,000 questions—only half the size of the EASA ATPL—with three answer choices (a, b, c) and a passing mark of 70%. Ultimately, the EASA ATPL exams involve answering 685 questions over the course of 20 hours and 45 minutes, while the FAA ATP consists of a single exam with 125 questions completed in 4 hours. Consequently, many view the EASA licenses as more prestigious due to the extensive effort and numerous exams required to pass. Additionally, the large number of questions results in a slower training process.

→ In Europe, the **number of airports** concentrated within a single training airspace is less than that found in a single U.S. state. For instance, Spain has 59 airports, whereas the state of Florida alone has 131 airports. Flying between airports in Florida involves much shorter distances. In my view, this simplifies training, especially when flying VFR in poor weather, as there are numerous nearby airports available for landing.

NOTE : I want to point out that some FAA certificate holders are allowed to fly in Europe as long as they are flying a U.S.-registered aircraft, since it is considered an American aircraft no matter where it is located.

- ✈ Conversely, the same rule holds for European-registered planes on American soil. This is also true for the number of schools. In the 1980s, there were 827,000 active FAA licenses, compared to 609,306 by the end of 2017. The primary reason for this is that over the past twenty years, more schools have been established in Europe and other regions. In the '80s and '90s, many individuals from Europe and other continents traveled to the U.S. for their studies. There are still more flight schools located in the U.S. than in Europe, and these American institutions have been training pilots for a longer period compared to some of the newer schools in Europe.

- ✈ In the U.S., pilot students have more flexible and lenient **rules**. For instance, when pursuing an ME rating, they are allowed to completely shut down and restart an engine. In contrast, in Europe, this is prohibited, and only an engine failure simulation at idle thrust is permitted.

- ✈ All FAA training is conducted in **English**. While many European schools offer EASA training in English, it isn't entirely in English. For instance, when I completed my PPL in Spain, my instructor was also a local, so we didn't always speak English. Additionally, a substantial amount of ATC communications is conducted in the local language rather than English.

- ✈ The toughest aspect of training under EASA is undoubtedly the 14 ATPL theory exams. On the other hand, the most difficult part of training under the FAA is passing the check ride, which involves a lengthy and challenging oral exam before the flight. During this oral exam, the student must explain any topic the examiner requests. If the student fails the oral exam, they cannot proceed to the check ride. Conversely, in Europe, the student completes a written test several days prior to the skills test (equivalent to the FAA's check ride, which is the practical part).

A few more acronym differences:

EASA	FAA
Skill test	Check ride
License	Certificate
ATPL	ATP
Mpl, modular and integrated training options	Modular training only
Mass and balance	Weight and balance
Pressure measurement in hectopascals	Pressure measurement in inches of mercury
Visibility in meters	Visibility in statue miles
Weight in kilograms	Weight in pounds
License is a paper folded into 4-6	License is a card type

FAA licenses are issued as "card type," while EASA licenses (UK CAA) are provided as "paper folded type."

EASA

	AGE	MIN. HOURS	ALLOWED TO BE PAID	MEDICAL CLASS	VALIDITY	NUMB OF THEORY EXAMS
LAPL	17	30	No	2nd	24m	9
PPL	17	45	No	2nd	24m	9
CPL	18	-	Yes	1st	n/a	13
ATPL	21	1500	Yes	1st	12m	14
FI	21	n/a	Yes	1st	3y	1
Full MPL	18	240	Yes	1st	Till unfreeze the atpl (1500h)	14 (atpls).

FAA part 61

	Age	Minimum hours	Allowed to get paid	Medical class type	Validity	Theory questions test
Sport pilots	17	20	No	3rd	*	40
Recreational pilot	17	30	No	3rd	*	40
Private	17	40	No	3rd	*	60
Commercial	18	250	Yes	2nd	*	100
CFI	18	n/a	Yes	3rd	*	100
Atp	21 and 23 unrest	1500 unrest	Yes	1st	*	100

*Validity of the certificates: All FAA pilots' certificates are valid as long as the medical by an approved FAA doctor is current.

- student pilot minimum age for balloons and gliders is 14
- student pilot certificate minimum age for others is 16

I've included a few examples of EASA ATPL and FAA ATP questions to give a general sense of their content.

EASA ATPL type

In a turn with speed brakes extended, roll control spoilers:
a) move on the down going wing only
b) move on the up going wing only
c) move up on the down going wing, down on the up going wing
d) move down on the down going wing, up on the up going wing

What is the effect on induced drag of weight and speed changes?
a) Induced drag decreases with decreasing speed, and induced drag decreases with increasing weight
b) Induced drag increases with decreasing speed, and induced drag increases with increasing weight
c) Induced drag decreases with Increasing speed, and Induced drag decreases with decreasing weight
d) Induced drag Increases with Increasing speed, and Induced drag Increases with decreasing weight

vs. an FAA ATP question:

When are inboard ailerons normally used?
a) High-speed flight only.
b) Low-speed flight only.
c) Low-speed and high-speed flight.

What is a purpose of flight spoilers?
a) Increase the camber of the wing.
b) Direct airflow over the top of the wing at high angles of attack.
c) Reduce lift without decreasing airspeed.

The correct answer is "c" on all four questions.

DECIDING TO STUDY ABROAD?

Deciding whether to pursue training overseas or remain in your home country can be challenging. This decision heavily relies on your personal circumstances and financial situation. If you are employed or have children, you will likely opt for training at a local school in your hometown. If you aren't tied down to anything, I highly recommend studying abroad due to its numerous advantages. It will challenge you to step outside your comfort zone, foster personal growth, rapidly enhance your English skills, offer an entirely new cultural experience, and, most excitingly, you'll make friends from across the globe!

13.4 Staying current

Keeping all your licenses up to date can be challenging. You must plan ahead for a medical appointment and reserve time with the school before retaking a skills test, possibly scheduling time with the simulator too. If you haven't flown recently, remembering all the procedures can be demanding.

Additionally to this, staying current costs money. If you're employed by an airline, they typically handle this cost, but if you're not, you'll have to pay for it out of your own pocket.

You can renew your licenses at any flight school. The flight school will assess whether you require refresher training, which might include some simulator sessions, before you take the airplane skills test.

Twice annually, airline pilots visit the simulator to maintain their qualifications—once for license renewal and once for line currency. We engage in exercises involving various abnormal scenarios and emergencies. Each year, we also undergo fire training and review all the onboard emergency equipment, as well as procedures such as evacuation and addressing smoke in the cabin.

Difference between revalidation and renewal

You are allowed to revalidate your rating if it hasn't expired yet, within three months before it does. Once the rating has expired, a renewal process will take place.

Distinction between proficiency and currency (specific to the FAA)

Currency and proficiency have similar definitions, and they do complement each other, but neither one is a replacement for the other. Being <u>current</u> under the Federal Aviation Regulations means that you have met the requirements to act as a pilot in command of an aircraft within a specific time period. Being <u>proficient</u> means, according to Webster's College Dictionary, "fully competent in any art, science, or subject." You can be current without being a proficient pilot, but if you are proficient, most likely, you have also met the currency requirements to get to that point.

To act as pilot in command of an aircraft, you need to have accomplished a flight review or one of the exceptions within the preceding 24 calendar months. To carry passengers, you need at least three takeoffs and three landings within the preceding 90 days. For night flights with passengers, the landings need to be a full stop at night, not touch and go. When the IFR environment is considered, at least six approaches need to have been performed and logged within the preceding six calendar months. There might be further requirements and exceptions to each of these situations, and we strongly recommend a thorough reading of the regulations to verify your requirements.

> ✏ **NOTE**: Ensure you renew your ratings well before they expire. Staying updated is as simple as reviewing the FARs relevant to your type of flying.

14 – CONVERTING LICENSES

14.1 Foreign license conversion to FAA

Why should you convert your licenses to FAA? One reason is that if you plan to fly an aircraft registered in the United States (N registered) anywhere in the world, you'll need FAA licenses. Alternatively, if you have the right to live and work in the U.S., you might want to obtain FAA licenses. There are numerous reasons for converting your licenses.

The procedure is straightforward, just involves some paperwork.

1. The first step will be to validate your foreign license. To do this, you must proceed to the FAA website under the search of "foreign license verification." This procedure takes around 45-90 days, and once it is verified, you will receive a "Verification Letter." From now on, <u>you have six months to start training</u> in the United States
2. Apply for a student visa (explained in detail in chapter 14.2.3)
3. Request TSA approval
4. Once you have your visa and your verification letter, then you can travel to the U.S.
5. Get an FAA Medical Class 1
6. Start your conversion training, which most institutions refer to as the ATP CTP certification training program.

i. Pass the ATP theoretical test (124 questions, 4 hours, and minimum pass grade of 70%)
ii. Pass the instrument rating theoretical test
iii. Receive instruction on a multi-engine airplane or simulator
iv. Pass the check ride (skills test/ practical) for ATP and ME-IR

This entire process is expected to last between a week and ten days.

The total cost of the conversion should be around 8000$, and if you are also converting a type rating like, for example, A320, then approximately 15.000$ and a period of 3 weeks to one month.

> ✎ **NOTE:** To utilize your FAA PPL privileges, you must always carry your original license, as the FAA PPL relies on it.

14.2 Foreign license conversion to EASA

Why should you convert your licenses to EASA? European airlines often hire pilots with minimal experience, sometimes as little as 200 hours. Converting your foreign license to a European one is more expensive and time-consuming because of the 14 ATPL exams required. Before beginning EASA training, ensure you have arranged any necessary student visas.

Future Pilots Checklist

The conversion divides into two categories:

a) >1500 hours obtaining a full ATPL license
b) <1500h obtaining a fATPL license (frozen)

a) Full ATPL

1- Get an ATPL ground course, either full ground school or online distance e-learning
2- Get an EASA Class 1 medical (to facilitate the process, ideally under the same civil authority, you will obtain your license) described in chapter 14.1.1
3- Get an English Language Proficiency minimum level 4 (described in chapter 13)
4- Do a skill test on your type rating aircraft or a full flight simulator
5- Apply for an EASA license to the Aviation Authorities

b) fATPL (frozen)

1- Get an ATPL ground course, either full ground school or online distance e-learning
2- Get an EASA Class 1 medical (to facilitate the process, ideally under the same civil authority, you will obtain your license) described in chapter 14.1.1
3- Get an English Language Proficiency minimum level 4 (described in chapter 13)
4- Take around 20-35 hours of flight training
5- Take a skill test for CPL + MEIR
6- Apply for an EASA license to the Aviation Authorities

Please note that you'll need to complete an MCC and JOC, or an MCC APS. The whole process can take anywhere from six months to a year, depending on the school, how quickly you finish the 14 ATPL exams, and the country where you'll obtain your license (for fATPL pilots). The prices range from 18.000 to 25.000 euros.

14.3 other conversions

For some individuals, pursuing flight training overseas is the only viable choice. Nonetheless, I believe it remains a worthwhile option to explore. This is particularly true if there are no reputable professional schools nearby or if English is not your native language. Even though all professional flight schools conduct their training in English, being immersed in a completely English-speaking environment offers a distinct advantage.

Why is training in Spain so popular across Europe? The answer lies in its favorable weather for about 80% of the year, along with more affordable accommodation and services compared to other European nations. Similarly, Florida in the U.S. is another hub for flight schools, offering competitive prices.

While rain and storms can occasionally pose difficulties in Florida, you can usually enjoy sunny weather for the majority of the year. Additionally, you have the opportunity to complete the training entirely in English.

In the U.S., there are schools where you can obtain both EASA and FAA licenses simultaneously. If you choose to study in the U.S., you'll need to apply for a visa via the U.S. embassy in your country.

One of the advantages of flying in the U.S. is the wide selection of airports available and the various instrument approaches you can practice. You also get the experience of taxiing behind large planes at major international airports. The aviation culture there is distinct as well.

> ✎ **NOTE:** To work in the U.S. under FAA licenses, you are required to have the right to live and work in the U.S.; hence you will need an American passport or a green card or under an F-1 visa. To work in Europe under EASA licenses you require a EEA / European passport with the unrestricted right to live and work in the EU.
>
> ✎ **NOTE:** Canadian TCCA licenses generally adhere to the same standards as the FAA. For details on converting your American FAA licenses to Canadian, visit the TCCA website.

15 – FLIGHT TRAINING

15.1 Tips

Flight training is meant to be both challenging and enjoyable. Reflecting on it now, I have wonderful memories. I met incredible people and still keep in contact with my instructors. However, it wasn't always blue skies.

I encountered challenges that ultimately strengthened my skills and boosted my confidence as a pilot. This section aims to equip you with an understanding of what to anticipate and how to navigate the obstacles you may encounter during flight training. Here are a few examples:

- ✈ **Bad instructing**: Not everyone is suited to be an instructor, and some are in the role simply to fulfill their required hours. I've witnessed some of my colleagues feeling extremely demotivated, anxious, and frightened before a flying lesson. This is something that should never occur. You should be thrilled to fly the plane and to practice the new maneuvers you've learned! There was a time when I had to fight back tears because an instructor was shouting at me for messing up. How can anyone learn effectively under such pressure?
  When training becomes negative, a student may eventually stop learning and progressing. Instructors

must recognize that each student is unique, with individual challenges and limitations. It is the instructor's responsibility to identify these issues and assist in overcoming them!

SOLUTION: You're the customer and you're paying a significant amount of money. Approach the head of training to discuss the problem and request a different instructor. It's that simple. Don't let the situation escalate. If you sense that the instructor is bringing you down rather than lifting you up, discuss it with them. If things don't improve, request a change of instructor.

- **Motion sickness:** During flying lessons, some individuals experience sickness and nausea, sometimes even vomiting. The aircraft is small, lacks air conditioning, and the steep turns combined with the G-forces can be overwhelming, especially if you're not accustomed to such motions. This discomfort can make it difficult to enjoy flying and keep up with the lesson.

SOLUTION: Motion sickness isn't very common among student pilots, affecting only about one in ten. If you're among those affected, consider taking motion sickness tablets before your lesson. Make sure not to fly on an empty stomach but also, avoid attending a flying lesson after consuming a large meal.

If you're in a place with high temperatures, request to have your lessons in the early morning or late evening when it's cooler. Once you start flying commercially, you can be confident that motion sickness won't be an issue anymore because large airplanes are more comfortable and stable. You won't be performing extreme maneuvers like 45-degree turns and stalls.

A colleague began feeling unwell about 40 minutes into the flight. His instructor observed this and initiated shorter adaptation flights starting at half an hour, gradually increasing the duration. This approach might be another effective solution.

- ✈ **Personal issues:** Personal issues can disrupt flight training. During my time at flight school, a fellow student experienced his parents' divorce. As a result, he was unable to finish the training at that time, although he later returned and completed it. He struggled to focus on his studies and enjoy the experience because his mind was elsewhere.

SOLUTION: Unexpected events can arise and impact the training. I suggest implementing a possible solution as soon as feasible. If the situation is more serious, such as the loss of a family member, consider taking some time off to mourn and recover.

If you find it hard to focus and proceed with your training, stop and discuss your circumstances with the school management. They'll likely be very empathetic, as this could occur to any of us at any time, even when we're working for a company.

- **Feeling scared**: One of my colleagues experienced a propeller strike during a solo landing. This incident scared her, and she decided not to pursue further flight training. Situations like this, or other events, can cause a loss of confidence in your ability to become a pilot.

SOLUTION: If a similar situation occurs during training, it's crucial to resume flying the very next day. It's like having a minor car accident; driving again might be tough and nerve-wracking, but the quicker you do it, the sooner you'll overcome the fear.

Things happen, we learn from them, and we move on! This is similar for airline pilots in the simulator; we make errors but strive not to let them affect us throughout the remainder of the session.

Don't allow an error or incident to derail your aspiration of becoming a pilot and disrupt your training.

- **Failing a test**: If you tend to be a perfectionist, you likely find it difficult to accept having a failed test on your training record. I've observed individuals become quite upset after failing certain tests, especially a flight test or a knowledge exam.

SOLUTION: Don't be too harsh on yourself. Not passing a test doesn't mean you've failed in your pilot training journey. Consider it a learning opportunity: Was I not adequately prepared? What should I focus on studying or practicing more?

What can I do better next time? Everyone has off days; maybe you felt unwell during the exam, or something else went wrong. Whatever it was, just let it go; things will definitely be better next time!

- **Struggling to follow up**: Flying requires a variety of skills, including the ability to multitask. You need to listen to the radio, monitor flight instruments, pilot the aircraft, and pay attention to your instructor all at once. This can often be overwhelming and challenging to manage everything properly.

SOLUTION: Prepare yourself very well in advance. Once you step inside the airplane, you already need to know what you have to do. Go repeatedly over the procedures in your mind, take a piece of paper and write them down until you know them by heart very well. Preparing in advance will make the flying lesson easier, and you will be able to concentrate on your instructors' explanations and demonstrations.

- **English language**: This can set the whole training backward, and you might end up behind your colleagues if you struggle to communicate and understand the lessons.

SOLUTION: Dedicate additional time to studying aviation terms you find challenging. Search for their meanings to ensure that when your instructor or ATC speaks to you, you can comprehend and reply appropriately. In your spare time, listen to www.liveatc.net.

- **Low on funds**: You had to take additional flying lessons, which you hadn't planned for when you signed up for flight training. Perhaps you encountered issues with your car or housing, and now your finances are insufficient to complete your schooling.

It's possible you're a spender and have exhausted all your funds for flight training on things like partying, dining out, costly accommodations, a new car, and so on. (Believe me, I've witnessed all of this!)

SOLUTION: Resolving this issue might be challenging. You may need to halt your training until you secure additional funds, or revisit your funding sources, such as the bank, your parents, or returning to work, among other options.

- **Losing the medical**: You are unwell or need to undergo medical treatment for some time, which prevents you from continuing with your current flight training. Preparation is the key to success in flight training

SOLUTION: You need to discuss your case with your AME. They will provide the best guidance on how to resume flying quickly. Don't worry that this might mean the end of your flying career.

FLIGHT SCHOOL TRAINING CHALLENGES	
DIFFICULTY	SOLUTION
Personal issues →	Take time to solve them or to heal
Motion sickness →	Take motion sickness tablets
Bad instructor →	Talk to him or her, or change
Got scared or lost confidence →	Get back to flying asap
Failing an exam →	Prepare well in advance
Struggling with English language →	Study aviation English on your free time
Running out of money →	Get back to the source if able

I have a colleague who underwent surgery. After the procedure, he needed three months to recover. Once all his medical tests were sorted out, he returned to flight training.

Altogether, he spent six months waiting for the surgery and then recovering. Communicate with your school administration. Let them know that you're working to resolve the situation and plan to return to your studies.

There will be days when it feels like everything is crashing down on you. Go over the day's lesson and allocate between two to four hours daily for ground school. If you're preparing for your European ATPLs, plan for additional study time.

<u>Focus on getting through the present day</u> and allow time to prepare for tomorrow's lesson. I cannot emphasize enough how important it is to show up to the class and to fly well prepared. If you don't believe me, you will see for yourself!

Flying is expensive, do not waste the money just because you have not prepared well in advance, or you were too busy yesterday at that pool party. In fact, I highly recommend setting aside some time each day for yourself to engage in activities that help you relax and de-stress. This could include listening to music, taking a run outside, watching a Netflix show after dinner, or swimming in a pool if you have access to one. Establish a routine by waking up and going to bed at the same time daily, rather than staying awake all night and sleeping in until lunchtime.

15.2 Equipment

As additional information and a guide for your pilot training, here is a list of equipment you'll need to succeed in your pilot training!

- ✈ <u>Headsets</u>: I recommend David Clark or Sennheiser to start training. Some schools allow you to use theirs, but it is best to have your own since this is quite personal. If you are looking for a professional

headset with noise cancellation, I recommend any model from Bose. It's excellent when spending eight to ten hours on a cockpit; the downside is that they are a bit pricey
- Kneeboard: It's helpful to have the plates or some paper available to write down clearances and other details like the ATIS.
- Fuel tester: A tiny container used to extract water from the base of the fuel tanks before the flight.
- Foggles: These glasses are designed for IR training. The lower section is clear, allowing visibility, but the upper portion is intentionally blurred to prevent looking outside.
- Chart board: a standard dinA4 board will be useful to hold your on-route charts folded
- Erasable color pens: to mark on the chart the route you will be doing
- e6b calculator: It serves various purposes! You'll need it for your written exams and for flying as well.
- Flight bag: to carry all your stuff inside
- Uniform: the school should either supply one for you or have a store where you can buy it
- Headlight: A regular flashlight can be useful, but there are times when you need your hands free to check engine oil levels at night or at dusk, or to examine approach plates or a plotted chart. Many sporting goods stores carry them. Many sports stores have them available
- Logbook: to keep track of your flying hours
- Calculator: If your school permits, you can also use your phone's calculator, which is handy for weight and balance calculations.
- Books: The school is supposed to supply them, but you could also buy extra training books that are useful.

Future Pilots Checklist

The total cost of all this equipment is roughly 1200 euros or dollars. If you're on a budget, consider purchasing second-hand equipment from graduate students who no longer need it. During my flight training, I bought some used items and sold others after completing my course.

16 – THE FUTURE OF AVIATION?

What lies ahead?

Will the demand for pilots remain high in the future?

What direction is aviation heading?

The rising demand for aviation jobs becomes more apparent each year because of:

- ✈ Growth in airport numbers as well as infrastructure, increasing the global carrying capacity
- ✈ The expanding middle class, particularly in Asia, is making air travel more affordable.
- ✈ A significant portion of the current pilots is approaching retirement age. This creates a substantial gap that needs to be filled with new, qualified professionals.
- ✈ Reduced fares increase demand (more low-cost companies)
- ✈ Airlines expanding and growing its fleet to meet air travel demands

According to a Boeing report*, in the next 20 years, there will be a demand of 800.000 civil aviation pilots, 769.000 aircraft maintenance technicians, and 914.000 flight attendants worldwide to supply the massive demand for air travel.

These figures encompass the civil, executive, and helicopter sectors.

Source : www.boeing.com/commercial/market/pilot-technician-outlook/

To provide you with more statistics and a sense of how intense the future demand for aviation jobs is, FlightRadar24 reports that at any given moment, there are between 8,000 and 20,000 airplanes in the sky. During the winter, particularly around Christmas, the number of departures is at its lowest. In contrast, during the summer months of July and August, the number reaches its peak, with over 800 departures an hour!

IATA reports also reflect on the increasing numbers of passengers and cargo transported yearly. In 2019, 4.5 billion passengers were flown, and it estimates by 2037 to 8.2 billons. China has surpassed both the U.S. and other countries in the Asia-Pacific region, such as Indonesia and Thailand, to become the largest market.

With the rise of low-cost airlines, people are traveling more frequently and efficiently than ever before. Weekend trips have become viable! Additionally, the availability of products has increased dramatically; even the apple you just ate may have traveled 5,000 miles to reach you.

17 – LAST WORDS FROM ME...

If you've reached this point, congratulations! Your dedication to becoming a pilot is evident, and it shows your persistence and eagerness to explore this fascinating and dynamic world.

With countless blogs and YouTube videos offering similar content, my goal for this book was to organize and consolidate all that information into a single guide to assist you in making informed decisions about your career and education. Some flight schools will tell you that some training is better than others and that a particular airline will want such and such. As knowledge is power, you'll approach the flight schools with a wealth of information. Plus, you can continually reference this guidebook throughout your training.

I know that the journey to becoming a pilot can be unclear and confusing, especially when figuring out where to start. That's why I've compiled a concise summary in a checklist format of nearly every chapter of this book, laid out step by step.

Show respect for this profession, as this role is no longer just about you. Wearing the uniform means you represent not only all pilots globally but also the airline you work for. Consider the implications if a pilot is spotted intoxicated at the terminal or consuming alcohol while in uniform (and yes, it has happened). This kind of conduct harms the reputation of other pilots who are dedicated and responsible in their profession. Passengers can quickly lose trust in us. Consider incidents like Germanwings; they instill a great deal of fear of flying in people. How can passengers be sure that a pilot won't commit a similar act? Handle your personal issues outside of work—seek professional help or take leave if needed. However, the moment you enter the flight deck, you must maintain a professional demeanor.

While training, pay close attention to your captain's and instructor's feedback, and critically evaluate their suggestions. Take notes on areas for improvement to help you develop and grow as a pilot. Leave your ego at the door before entering the flight deck. There's no room for arrogance in aviation, as it has previously led to the loss of many lives. You will be in charge of over 200+ passengers who entrust you with their lives and souls, relying on you to transport them safely to their destination, even though they have never seen or met you.

Lastly, take your training seriously but also enjoy it! The flights you will do with your instructor at the school will be so much fun! The first time you fly solo and land by yourself is a day that you will never in your life, forget. The moment you get a call or an email from a company informing you that you've been chosen, it's the second peak of excitement.

And with that will come the most amazing sunsets and sunrises from the highest office in the world.

The walk is not easy, but certainly, the view makes worth every effort.

Remember, you are our future pilot.

Future Pilots Checklist

See you in the skies!

Captain Prat

18 – YOUR FINAL CHECKLIST

As you advance in your journey to becoming a pilot, follow the checklist below in the correct sequence to assist you throughout the process.

FUTURE PILOTS CHECKLIST		
1	Do you have funds?	☐
2	Pick up a training location	☐
3	Extensive online school research done	☐
4	Visit the top 3 schools?	☐
5	School paperwork ready?	☐
6	Medical certificate completed	☐
7	PPL / Private certificate achieved	☐
8	IR / Instruments completed	☐
9	Time building completed	☐
10	ATP / ATPLs finished	☐
11	Commercial and ME completed	☐
12	Got your ELP (EASA only)	☐
13	MCC (EASA only) / CFI (FAA)	☐

Be sure to follow us on our social media accounts and to subscribe to the newsletter!

@futurepilotschecklist

www.futurepilotschecklist.com

19 - SUCCES STORIES

The personal stories in this collection differ a lot in between them, showcasing how individuals from various backgrounds, ages, and financial challenges pursued their commercial pilot licenses.

All these stories were written and collected in 2020 when the book was first written and published.

I hope they inspire you!

19.1. Carlos´s age success story

"Imagine for a moment this world: a world in which every day you wake up unmotivated, without wanting anything, without a goal to achieve, without knowing where you are or where you are heading, without being able to imagine what the future will bring you and without even wanting to make plans in the short, medium and much less, long term, because the uncertainty about your professional future is simply so deep, that you are mired in one of the greatest fears ever experienced by you: the fear of leaving your comfort zone, of growing up, of overcoming yourself, the fear of change, of chasing your dreams.

This was my reality for many years.

From a young age, disgrace hit my family several times with traumatic deaths, which, among other things, led me to choose an unwanted career path. However, it was a good path, with relative success and good job promotions, based on effort, perseverance, and good work, but it was far from my dream path.

Once you get caught by aviation, you never get it out of your soul. I was captivated around 1988 by a flight from MAD to NYC in a TWA 747. At that time, boarding by an airbridge was not something very common and, when the passenger bus left you on the platform next to an iron mass of enormous dimensions like the queen of the skies has, myself measuring in between 130 and 150 cm, you can imagine how different was the proportion of sizes between the two, and how much perplexity that marvelous piece of aeronautical engineering could generate in the fanciful imagination of a ten-year-old boy. Suppose you add that the flight attendant allowed all the children on board to enter into the cockpit back then. In that case, it is during moments like that, when I saw that festival of analog clocks, buttons, and indicators, where some of the sickest passions that exist are forged, the love for flying.

The power of the mind is curious. I can still remember the smell of air freshener used on board that Jumbo, a mixture of lemon and orange blossom, memories burned into your existence of experiences that have marked you forever.

These powers, which allow you to recognize the smell of lemon and orange blossom while walking through a park, and transport you back aboard that TWA Boeing twenty-five years ago, are the same ones that allow you to do "a click" one day and help you fight to get out of your comfort zone, to propose things you didn't even imagine you could do, to draw strength from where you didn't have, and to pursue your dream so hard until you achieve it. Those are the powers that we all have and that we only have to search within ourselves until we find them. In my case, they emerged in 2010, and although there were several external factors that I will not go into detail, and that helped me remove them, I simply took advantage of them to achieve my goal.

I have always been a hard-working person; from eighteen to thirty-two, I worked in a technical construction office. Although I was still a bad student, the effort and determination that characterize me made me progress step by step in the different companies in which I worked until I reached a respectable level of responsibility and a fair salary. My work offered me a lot of freedom of schedule and organization, I had to deal with many people, and for someone like me being very outgoing, this is something fundamental. But that did not fill me, I continued to feel an emptiness in me. There was still something that did not fit, a piece of the puzzle was still missing.

So after a series of anxiety attacks that lasted almost six months and that woke me up in the middle of the night, forcing me to wake up to make sure that I was still alive, I understood that I had to do something and take the reins to redirect my work life and thus at the same time, achieve greater general well-being.

At the age of thirty-two, I began to gather information about the requirements to become a pilot, and although the necessary investment can scare anyone, I must say that if you want, it is possible. There are always several resources that will allow you to face it, even if they involve a lot of effort.

In my case, I financed the PPL with a series of coins and small gold bars, which my family gave me during the first eighteen years of life, the distinctive gifts of baptism, birthdays, etc. My family had been able to turn these gifts very intelligently into those little investments that never lose value, quite the opposite.

From then on, everything went smoothly; step by step, I was doing my pilot training in a modular way, paying on my path and basically renouncing both my wife and I to make any kind of trip or whim to finance my dream. While taking my theoretical ATPL course, I had the great opportunity to start working as a sales manager, informing and selling the courses taught at the flight school. I knew that this was my gateway to a future job as a flight instructor, so I did not miss the opportunity. In life, opportunities arise from time to time. You just have to be able to see them, not let the challenge scare you, and hold onto them without letting go.

For three and a half years, I combined my work activity with flight training at the same school. There were hard times in which the acquisition of clients was not easy since we came out of the financial crisis of 2008, and in which I had to work many hours to be able to combine both activities and obtain acceptable results at work, that would allow me to continue paying for all my pilot training.

Future Pilots Checklist

Once I finished my commercial, multi-engine, instruments, and flight instructor courses, finally, at the age of 37, I started working in aviation as an instructor—a significant milestone, a great responsibility, a great emotion, a great fear too—.

I managed to accumulate around 1700 hours of flight by working another three years as an instructor. They gave me confidence, security, caution, experience, knowledge, and courage, but my entry ticket to the airline industry was the most important thing for me.

I have always thought of the profession of the instructor should be reserved exclusively for people with a lot of experience in flight, with a lot of knowledge and background, people being able to transmit it comfortably and safely to their students, instead of a fast way to add flight hours to qualify you for future jobs in aviation. But right now, the system works like this, and I think that's another discussion that cannot be addressed here. I just took advantage of it.

So, at the age of 40, and after being immersed in more expenses doing screenings in several airlines, I finally passed the assessment to join Volotea, an airline based in Venice, flying the magnificent Boeing 717. That first airline experience allowed me to make the leap to the airline that I had pursued from the beginning Vueling. It allowed me to return home with my family, with my wife and son. Nowadays, being a pilot and working based in the city that one comes from, is a privilege available to few, so I simply cannot ask for anything more to be happy.

I am still paying the cost of my training. Having completed two type ratings, a price that will still take a while to pay back, but today, I can tell you that if you want, it is possible, it can be achieved and, that it is never too late to propose a radical change in your life and pursue your desire to fly. No matter how old you are, if you seek it hard and put in a lot of effort, you can reach the skies.

My name is Carlo della Rocca, I am 42 years old (2020), and I am currently a First Officer flying an Airbus 319, 320, and 321 all over Europe: I encourage you to pursue and achieve your dream.

Carlo della Rocca"

Carlo and I met in the flight school back in 2012; when he was the public relations responsible, and I was doing my PPL. Now, I hear Carlo's voice many times on the radio while we are both working. It still thrills me so much, knowing where we stood eight years ago! Thank you very much for sharing!

19.2. Joanne's corporate success story

"Wake up Joanne, it's time to go on holiday" Up I jump, "yes, we get to go on an airplane!" it was 4 am, and my dad was waking my brother and me up to go on our family holiday. I was six years old, and I was more excited about the airport and airplane than the pool and beach!

Future Pilots Checklist

Fast forward 12 years, I got a job as a ground staff agent at London Heathrow airport, checking in and boarding passengers. I managed to afford a few flying lessons at that time, but I struggled to keep it up with it being so expensive and with my airport ground staff job. I needed to figure something else out. I saw an advertisement for Emirates airlines looking for cabin crew: tax-free money, life in Dubai, travel around the world, but more importantly, I would be able to work on my beloved airplanes.

At any given chance, I'd take my breaks in the flight deck, asking lots of questions and sometimes even getting to sit in the on it for landing! I lived in Dubai for 5 years, I worked hard and loved my job, but I knew I wanted to fly the plane rather than working down at the back. After saving a lot, I looked into different flight school locations: South Africa and the United States were the most appealing to me, the weather and the cost of flying were far more suitable than in the Uk. But I wanted to get my EASA license, so I needed a European flight school.
Eventually, I found a European flight training school in Fort Pierce, Florida. I decided to give up working for Emirates and put everything for flight training.

I really enjoyed the flight school. I was sharing a house (school provided accommodation) with two other girls who became good friends of mine. I did six months of ATPL ground school, and in between, I did my hour building, as I needed 100hrs solo before I could get my CPL. This was the fun part; I would often get up early and fly along the coast at sunrise, the skies were calm and quiet, and I got the chance to enjoy and love what I was doing. During some flight lessons, I would

put myself under so much pressure to do an excellent job to the point that I would forget why I was doing it, and every lesson would feel like a test. I almost lost my love for flying, as I was getting through each lesson and not enjoying it. I did my cross country flying to different states all over the U.S. I stayed at local airfields doing touch and goes; we would take a few planes and fly to Key West for the day. Flying in Florida was truly unique!
I also was working on my instrument rating, which looking back now; it was the most valuable training I received.

Once all exams were complete, and I had my 100hrs solo, I did my CPL in a multi-engine plane, tried to kill two birds with one shot, and save the cost of doing the multi-engine and CPL together.

I left the U.S. with EASA CPL/ME rating with an FAA IR. I needed to convert the FAA IR to EASA back in the UK. So I found the most welcoming flight school in Exeter, and I did my ME/IR there in a BE76.

Once everything was completed, it was time to start applying for a job. I knew this wasn't going to be easy, being a low hour pilot in such a competitive industry.
I happened to be at the right place at the right time (this happens a lot in aviation!), and I meet the duty flight officer of the private jet company I work at now; after having completed a sim check and some interviews, I was offered a job!
I could not believe my luck! I was flying Learjet with just over 300hrs. The Learjet simulator training was intense but very enjoyable; I was determined to know everything about the

plane I was going to be flying. I worked hard to pass all my exams, and I turned up for the simulator sessions very well prepared.

Once all passed, line training was the next hurdle to overcome, and with thanks to the wonderful team of line trainers, I got through the line training and was signed off.

The corporate world of flying is perfect for me; I get to fly jets and still have that one-on-one interaction with our clients, which I thoroughly enjoy. We have a schedule of six days on and three days off, with the option of working our days off if needed.

During the summer, the six days on are almost a given that we will be flying, but we can spend a lot of time waiting around at home on standby during the winter.

The destinations we fly to are my favorite part of flying. We fly to many new destinations, and some can be challenging with deep valleys such as Innsbruck in Austria and Chambery in France. But again, we train in the sim for notable airports like those.

We can spend 3 to 4 days in Lapland, Finland, having dropped off our clients and it works out better to wait for them to fly them back than to go back to London and fly up days later. So, we often get some long layovers in some incredible countries!

We do not have a flight attendant on board unless requested by the broker, so it is either the captain or myself who will go back and serve the clients, but with my previous cabin crew experience, I enjoy this.

Since I have been flying as a pilot, I have been very happy, and all the hard work and challenges I faced along the way have paid off and made me the pilot I am today.

No two days are the same, and I will always stay humble and learn something new every day.

Joanne Jobson

I met Joanne at the flight school in Florida back in 2013, to find out she had also been a flight attendant in Emirates like me! It is always fascinating to hear how different and exciting her life is as a corporate pilot. Thank you for sharing your story!

19.3. Piotr´s training while working as flight attendant success story

My name is Piotr; I was born in 1985 in Szczecin, a town in Poland. Since I was a very young child, I always had a strong interest in aviation.

By 2006, at the age of 21, I started working as a cabin crew for an airline based in Germany. After giving it a lot of thought, by 2014, I followed my childhood dream, and four years later, I became a professional aviator. That same year, which was a lucky one for me, I also became a father for the first time.

My wife, who was also a cabin crew like me, wanted to stay for the last four months of her pregnancy in her country home town, which is Bilbao in the north of Spain. During that time, I had taken time off from my cabin crew job to be with her and, finally, to start an integrated pilot training course after many years of working and saving a lot of money. I found a pilot school nearby to the place where we were staying, in Bilbao. After five months of training, I returned to work in Germany since my time off came to an end.

From the moment I was back to work full time, the pilot training became much more challenging to complete. One year had passed since I started when I decided to move out from Germany to Spain. It did not make any sense to be in Germany when my wife and child were in Spain, as well as my flight school.

During this period, my school had changed its location to Leon, about three and a half hours' drive from Bilbao. Back then, I started working full time at Barcelona airport as a cabin crew while I continued my flight training. As I split in between working in one city and taking my training in another different town, it was not easy. I was a new father with lots of lack of sleep, while at the same time, I was preparing for the ATPL exams and having a full-time job as a cabin crew. That meant for me working five days in a row with alternate schedules, either earlies or late shifts, and then only three days off to fulfill my training at the same time as taking care of my family.

Eventually, I completed all the 14 exams. I managed to do it by flying on every day off from work to Leon, 700km from where I lived, most of the time I traveled by plane using staff travel tickets to Valladolid, from there I drove my scooter about an hour and a half to the school. Every single day at the school, I used it to fly and achieve a CPL license. To complete all necessary hours and time building, it took me about seven months of intensive training, combined with work, traveling, and family! By 2018, I had become a commercial pilot. In 2019, I upgraded my license with the FI qualification, which I have finished in one school close to Barcelona this time, just to avoid traveling and be close to my job and to my family, which is now in Barcelona with me.

But to end my story, I have been lucky for the second time this year 2020, I have welcomed my second son and, after completing my FI training, I managed to continue working as cabin crew part-time and as a flight instructor. Even though the flight school where I work is in Burgos, I am incredibly grateful and privileged to have found a job as a pilot already. At the same time, I never stopped working in my cabin crew job since 2006, almost 15 years ago. I had two kids in the middle of the process. My message to you is that if you want to become a pilot and have a supportive family as my wife has been, anything is possible. I got my first job as a pilot in 2020 at the age of 35. I financed the training with hard work and with lots of savings from my cabin crew job. There are no excuses. You can work and have a family while you become a commercial pilot. Of course, you will not have almost any free time or holidays for some short period, but the sacrifice is worth in the end!

At this moment, I am waiting for this Covid19 crisis to pass and start working as an airline pilot in the next two to three years. By then, I will have 1000 or more hours as an FI.

I hope my story encourages at least some of you to see that with some sacrifice, you can do it too. Even find a job in difficult times!

Piotr Chrusciel

Piotr and I are colleagues at the airline, and I've had the privilege of flying with him. He embodies hard work and discipline, proving that dreams can come true if you strive for them. I wish you many safe flights and hope to share the flight deck with you soon! Thank you for sharing your journey.

19.4. Andrea's MPL success story

I only know one person to have an MPL license since it is not a very common program; I asked my friend Andrea to tell us about his journey.

Keep in mind that the airline is the one who sets the guidelines for the training program with the school that has the partnership with. This means that another airline with an MPL program might have different payment options, different bonds, etc.

I was incredibly lucky and fortunate to have been one of the 30 successful cadets out of 600 aspirants selected for the MPL program of a very well-known Italian airline.

I had previously been doing flight training in the United States, where I had accomplished a CPL. When I returned to Italy a few months later, I saw this incredible opportunity, and I did not hesitate to apply as quickly as possible!

The first step was a complicated selection process consisting of three phases: phase one was done online. I had to do basic mathematics and English and Italian language tests. Once successful, I progressed to phase two. Here is where I realized how many people applied for this program and that the competition to get my spot on the airline was going to be very tough! I traveled to Rome and checked in my hotel, and the next day, I took an aptitude test on a computer, more mathematics, physics, and another English test. Progressing to the final and third phase, you could see the number of people reduced compared to the second phase's earlier days. I had prepared phase one and two very well. Still, now it was all up to a group interview and a quick chat with a psychologist to evaluate the candidates' personalities like any other airline interview.

Now I had played all my cards, and all I could do was go back to my home and cross my fingers, waiting for the response. I had a good feeling I gave it all on the three phases of the assessment, but still, the days were very long waiting.

A couple of weeks after the last interview, I received a call from the airline, and my heart was racing fast as I was answering it. Everything I worked hard for had finally paid off. I was a successful candidate, and I had a first officer position within the company!

But not everything was sun shines and blue skies yet. I still had to achieve a Class 1 medical and pass all the training in acceptable standards and behavior to have that first officer offer. Another important thing to sort out before training was the financing of the program. The program had a total cost of 130.000€, out of which 75.000€ was deducted from our salary once we completed the training, and started operating on the line. The remaining 55.000€ was taken care of by the airline sponsoring the cadetship. Of course, with a bond to remain in the company for three years, the amount had to be paid back if a cadet was not completing this time. The airline gave us all sorts of facilities for the loan since they had agreements with individual banks.

Once all this was sorted, I was ready to begin training! I was back into a flight school again, meeting the other students full of excitement to get started! My colleagues were all between the ages of 18 and 28. They came from very different backgrounds, some had pilot licenses already like me, some were coming straight from high school, and others from different working backgrounds.

The first step was ground school, the most feared 14 ATPL knowledge exams. And I am glad that was the first part; once completed, we could focus on the fun of flying! We began the initial training like learning to fly and basic maneuvers on a multi-engine DA42. Only a few flights are done on a single-engine Cessna 172 for the solo flight, with a total of 8 hours. We were progressing to a complex simulator that would equal the MCC + JOC training—followed by an exciting UPRT training made onboard the powerful Extra 300.

The type rating was on the Embraer 170 family, which is the aircraft that we are currently flying in the company.

After type rating, 6 takeoffs and landings are done before begin line training already with passengers. We received the license after base training since the MPL program was now completed. This license is valid until unfreezing your ATPL license at 1500h, along with another skill test. I can proudly say I accomplished mine last April.

It took us two years to complete this MPL program, leaving with roughly 300 hours, including real aircraft time and simulator time, to jump on the right seat on the passenger jet.

This program does really takes you from having 0 experience and knowledge in aviation to be a competent airline first officer. The main goal of it is to join an airline, so by doing an MPL license, you do not get a PPL, CPL, or MCC and JOC. An MPL encompasses all of it in one. Unlike other licenses, most flight training is done on the airliner's simulator, in which you will be flying, leaving only 90 hours on the basic piston aircraft training. And from the beginning, you will be working as a team, even onboard small planes. It is only valid for multi-crew airplanes, meaning an MPL holder cannot fly a single pilot aircraft as it would be, for example, a Cessna Caravan* unless he or she had a single pilot rating.

Like any other pilot license, revalidation is done yearly at a simulator training center along with line checks at the airline.

To conclude my experience doing an MPL license, I must say that although I had a CPL before starting, I have learned a lot *by doing it. I have expanded my knowledge in aviation, and the best part of it, I have secured an airline job. The fact the training is straight forward direct to the airline makes it very intense with a very steep learning curve. With this, I want to say that this might not be for everyone, but If you are starting one, brace yourself because you are about to get your brains heavily squeezed onto 24 hours learning! It is expected from you to comply and follow the program thoroughly with high standards and good grades!*

Andrea Cossu

Andrea now happily flies the Embraer 175 and 190 soaring European skies. Thank you for sharing with us your MPL journey and wishing you safe flights!

20- FREQUENT Q&A

1- How much time is typically needed to secure a job after completing flight school?

Typically, it ranges from 6 months to 2 years, depending on the current demand for pilots.

2- When is the ideal time to begin flight training?

If you can afford a PPL, feel free to begin now. If you can't, determine how you'll finance it first.

3- Can you convert an FFA license to an EASA?

Yes, as it is explained in chapter 15.2.

4- What are the steps to obtaining a bank loan?

Consult various banks and inquire whether the flight school has a partnership with any particular bank for financing flight training. Consider exploring options through a pilot union. A solution is out there.

5- How can you manage anxiety?

Study thoroughly in advance. That's the best approach to handle it. If you're not prepared for an exam, postpone it. Anxiety and flying are a bad combination.

6- Tips to prepare for flight training

Whenever your colleagues take a flying lesson that you'll soon be doing, try to join them in the plane as a "jump seat." I found it beneficial to learn alongside my fellow classmates.

7- Ways to cope with failing a test

You need to recognize what occurred: were you anxious because you didn't study sufficiently? Was there a lack of understanding of certain questions? Perhaps you didn't get enough sleep? Whatever the case, you can prepare better to ensure you pass next time. Remember, failing an exam isn't the end of the world.

8- What is the minimum age requirement to become a pilot?

18 years old for a commercial and 21 for an ATPL / ATP.

9- What happens if a female pilot gets pregnant?

Once she becomes aware, she should contact the AME and the company. Generally, if she feels well, she can fly until the 26th week.

10- Should flight students consider purchasing a tablet?

Not necessarily, unless it's part of the school equipment or the student believes it will aid them during training. Some useful navigation apps can be advantageous for time building.

11- How difficult is it to learn aviation in English?

It is not difficult. You just need to get familiar with the topics and abbreviations. I recommend you have an adequate level of English before starting flight training.

12- Do certain airlines provide type ratings?

Certainly, most major airlines offer the type rating, but it always comes with a bond!

13- How many months does it take to get a commercial license?

The duration ranges from at least three weeks to six weeks, depending on the schedules of both the student and instructor, as well as the weather conditions.

14- Is surgery allowed for visual defects?

Always consult your Aviation Medical Examiner (AME) before undergoing surgery; it shouldn't be an issue. Many pilots, myself included, have had lasik surgery.

15- Do you fly every day the same route?

Pilots typically have varying schedules with diverse routes and destinations. However, those working for a small airline and stationed in remote areas might find themselves flying the same routes repeatedly.

16- How to study for exams

I typically condense the lesson into a two or three-page summary highlighting the key points. I then repeatedly review it until I can nearly recite it from memory. For lessons involving calculations, I simply practice them extensively.

17- Recommended learning textbooks

A great book is "Pilot's Handbook of Aeronautical Knowledge" by the FAA. I suggest checking out some

YouTube channels packed with great information: "Mentour Pilot" and "Captain Joe."

18- A common question is, what is the difference between ATP/ATPL and a commercial license? What opportunities do each of these provide?

A commercial pilot can fly any aircraft that is not involved in commercial air transport. Examples of jobs exclusively for commercial pilots include being a flight instructor (holding a CFI license), towing banners, conducting sight-seeing tours, and performing aerial photography. To operate an airliner, a pilot needs an ATP/ATPL license.

19- Will converting my license to FAA or EASA result in losing my original one?

You won't lose your initial license; instead, you'll acquire an additional one. This means you'll have both your original license and either the EASA or the FAA license.

20- Do you always fly with the same crew?

Typically, pilot rosters change regularly. At larger bases, you're less likely to fly with the same individual repeatedly. However, at smaller bases, you may frequently find yourself with the same team. Executive pilots tend to fly more frequently with a consistent crew.

21- How long did it take you to find a job after completing your training?

It largely varies based on the circumstances, but typically it ranges from six months to two years. In 2015, I managed to secure a pilot job in just six months.

22- Is it possible for you to choose between flying an Airbus or a Boeing?

A pilot rarely has the opportunity to choose. Generally, when you get a pilot job, you fly the type of aircraft the company uses. However, if you opt for a "pay-to-fly" program, you can select the aircraft type rating you want to fly.

23- Can an Airbus pilot fly a Boeing?

No, unless the person is certified to operate an Airbus. These are distinct aircraft with unique systems and operations.

24- I am 40 years old, am I too old?

No, you are not. You still have 25 years until reaching retirement age.

25- What level of maths and physics do I need?

I'd say the equivalent of a high school diploma.

26- Is only the Captain allowed to land, or First Officers can do it too?

Both the Captain and FO do land the airplane. I was often asked whether I, as a co-pilot, ever get to do this. Some people assume First Officers never do landings, but who would do it if the Captain became incapacitated?

27- When do you become a Captain?

The number of hours needed to be promoted to Captain varies based on the airline's requirements, typically ranging from 3,000 to 5,000 hours.

28- What is a flight schedule like?

The schedule consists of a mix of days off, early shifts, night shifts, or late shifts with layovers. It varies based on your workplace and can change slightly. My schedule is fixed: five days on early shift, four days off, followed by five days on late shift, along with some standby days. I always manage to sleep at home because I never have layovers. However, in cargo airlines and executive aviation, you might be away for several days or even a week.

29- Is it difficult to bring the plane down safely?

A commonly asked question, landing a plane is just skills gained over experience. However, it can become more difficult due to weather conditions such as strong crosswinds.

30- What is the most challenging aspect of pilot training?

Two challenges: securing funding for pilot training and landing a job without prior experience.

31- Which subject did you find most difficult?

In my opinion, principles of flight. However, many people find it challenging to grasp topics such as flight planning, meteorology and performance.

32- Can you choose your flight destinations?

No, the schedule is random, although you have the chance to bid for flights you want, or you can always swap with a colleague.

33- How do standbys work?

There are two categories of standby: home standby and airport standby. Home standby typically spans 12 hours, during which you must remain contactable by phone and be able to arrive at the airport within an hour of receiving a call.

Airport standbys are shorter, but you need to pack your suitcase with both winter and summer clothing, as well as enough items for a few days, because you never know where you might be headed, especially with long-haul airlines.

34- What is the difference between ICAO, IATA, JAA and JAR?

ICAO: (International Civil Aviation Organization) A United Nations agency, with the purpose of managing the administration and governance of the Convention on International Civil Aviation.

IATA: (International Air Transport Association) A trade association of 290 airlines (82%) with the purpose of strengthening the business within the industry by improving the safety, security, airport network, environment, etc.

<u>JAA:</u> (Joint Aviation Authorities) It was an associated body of the EASA representing the *civil aviation* regulatory *authorities* of the several European States who had agreed to co-operate in developing and implementing common safety regulatory standards and procedures. *

Source: *Wikipedia

<u>JAR:</u> (Joint Aviation Requirements) was a set of common comprehensive and detailed aviation requirements issued by the Joint Aviation Authorities, intended to minimize Type Certification problems on joint ventures and facilitate the export and import of aviation products. *

35- Can I work in the U.S after completing training there?

No, unless you have a residency or a green card, any airline sponsorship, or you obtained an F1 visa.

36- What if I aspire to become a helicopter pilot?

To become a helicopter pilot, the procedure is similar to the airline pilot. You would still need to complete the entire training for a helicopter license; any flight hours you have logged on an airplane do not qualify you to pilot a helicopter.

21 - GLOSSARY

AA American Airlines

AGK Aviation General Knowledge (ATPL subject)

AME Aeronautical Medical Examiner (Doctor)

ANAC National Civil Aviation Agency (Brazil)

ATC Air Traffic Control

ATO Approved Training Organization (Flight School)

ATP Airline Transport Pilot

ATP CTP Airline Transport Pilot Conversion Training Program

ATPL Airline Transport Pilot License

BMI Body Mass Index

CAA Civil Aviation Authority

CAAC Civil Aviation Administration of China

CAE Canadian Aviation Electronics

CASA Civil Aviation Safety Authority (Australia)

CFI Certified Flight Instructor (FAA)

CFIT Controlled Flight Into Terrain

Future Pilots Checklist

CPL	Commercial Pilot License
CRJ	Canadair Regional Jet
CRM	Crew Resource Management
CV	Curriculum Vitae
dB	Decibels (unit to measure sound)
DGAC	Directorate General of Civil Aviation (India)
DHL	Postal and logistics company
DME	Distance Measurement Equipment
DPE	Designated Pilot Examiner
EASA	European Union Aviation Safety Agency
ECG	Electrocardiogram
EEA	European Economic Area
ELP	English Language Proficiency
EU	European Union, Europe
FAA	Federal Aviation Administration (United States)
FAR	Federal Aviation Regulations (under FAA)
FBO	Fixed-based operator (Airport Services)
FedEx	Federal Express (Postal and logistics company)
FI	Flight Instructor

GPS	Global Positioning System
Hz	Hertz (unit to measure frequency)
IAA	Irish Aviation Authority
IATA	International Air Transport Association
ICAO	International Civil Aviation Organization
IFR	Instrument Flight Rules
ILS	Instrument Landing System
IMC	Instrument Meteorological Conditions
IOE	Initial Operating Experience
IR	Instrument Rating
ISWAP	International Society of Women Airline Pilots
JAA	Joint Aviation Authorities
JAR	Joint Aviation Requirements
JOC	Jet Orientation Course
LAPL	Light Aircraft Pilot License
LCA	Line Check Airman
LOT	Polish Flag Air Carrier
MCC	Multi Crew Cooperation
MCC APS	Multi Crew Cooperation Airline Pilot Standards

MDA	Minimum Decent Altitude
ME	Multi Engine
MEI	Multi Engine Instructor
MPL	Multi Pilot License
NASA	National Aeronautics and Space Administration
NDB	Non-directional Beacon
NR	Night Rating
PIC	Pilot in Command
PPL	Private Pilot License
RAF	Royal Air Force
SE	Single Engine
SEP	Single Engine Piston
SEVIS	Student and Exchange Visitor Information System
SFI	Synthetic Flight Instructor
SID	Standard Instrument Departure
SOP	Standard Operating Procedures
STAR	Standard Terminal Arrival Route
TCCA	Transport Canada Civil Aviation
TOC	Top of Climb

TOD   Top of Decent

TR   Type Rating

TRI   Type Rating Instructor

TSA   Transport Security Administration (U.S.)

UK   United Kingdom

UPRT Upset Prevention and Recovery training

U.S. United States

VFR   Visual flight rules

VS   Versus

WW1   World War one

WW2   World War Two

22 - ADDED DOCUMENTS
SIDS, STARS, APP, TAXI DIAGRAMS

AMSTERDAM AIRPORT DIAGRAM

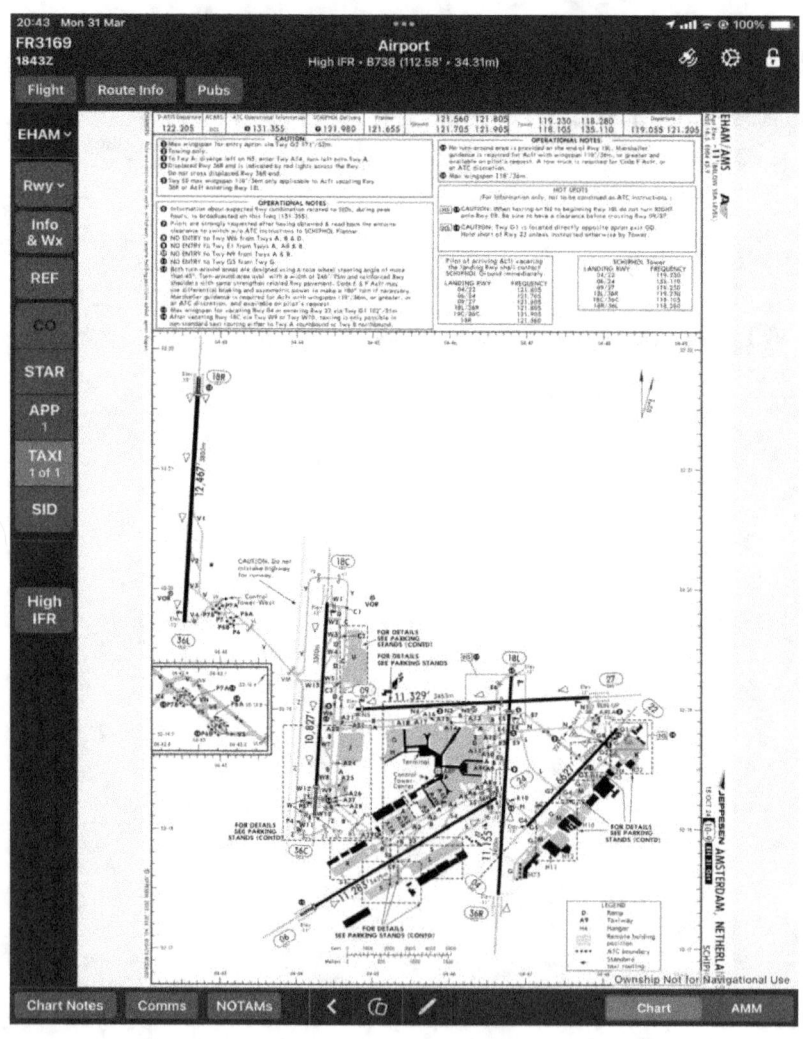

AMSTERDAM SID KUDAD 2G from runway 22 & 2N from runway 09

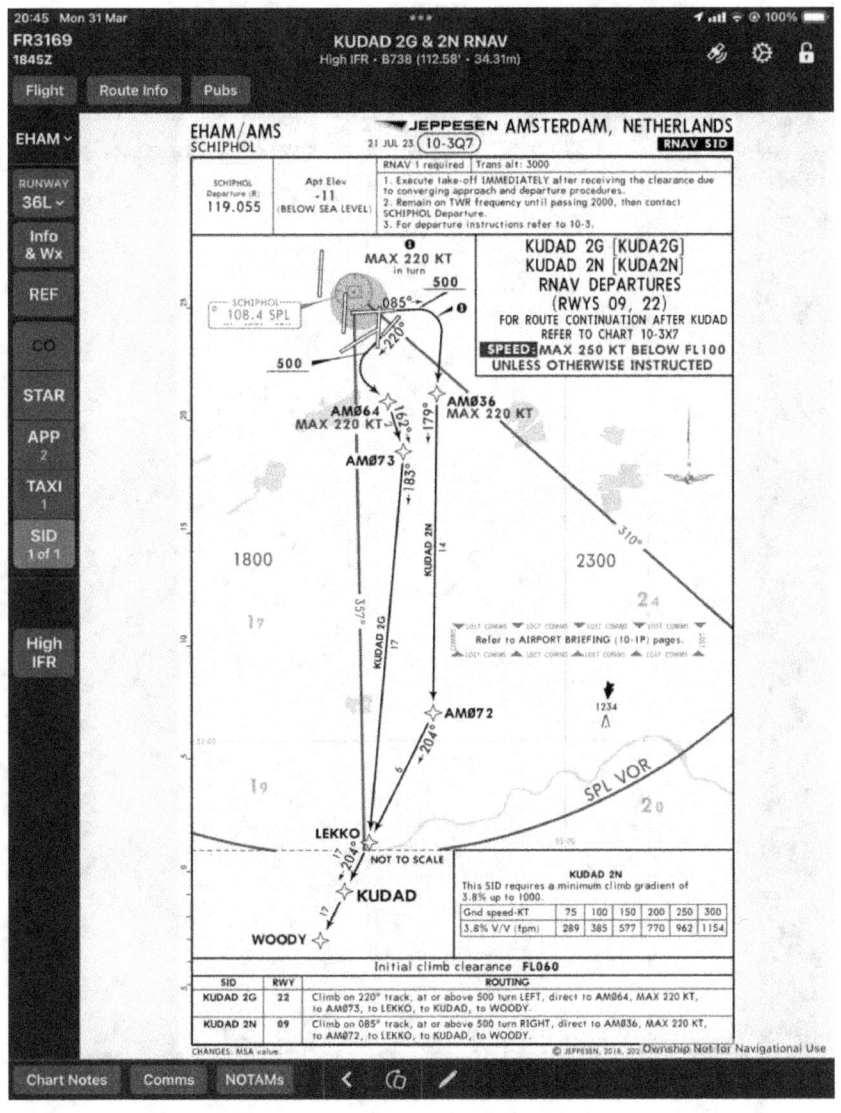

AGADIR (MOROCCO) HIGH TERRAIN AREA CHART

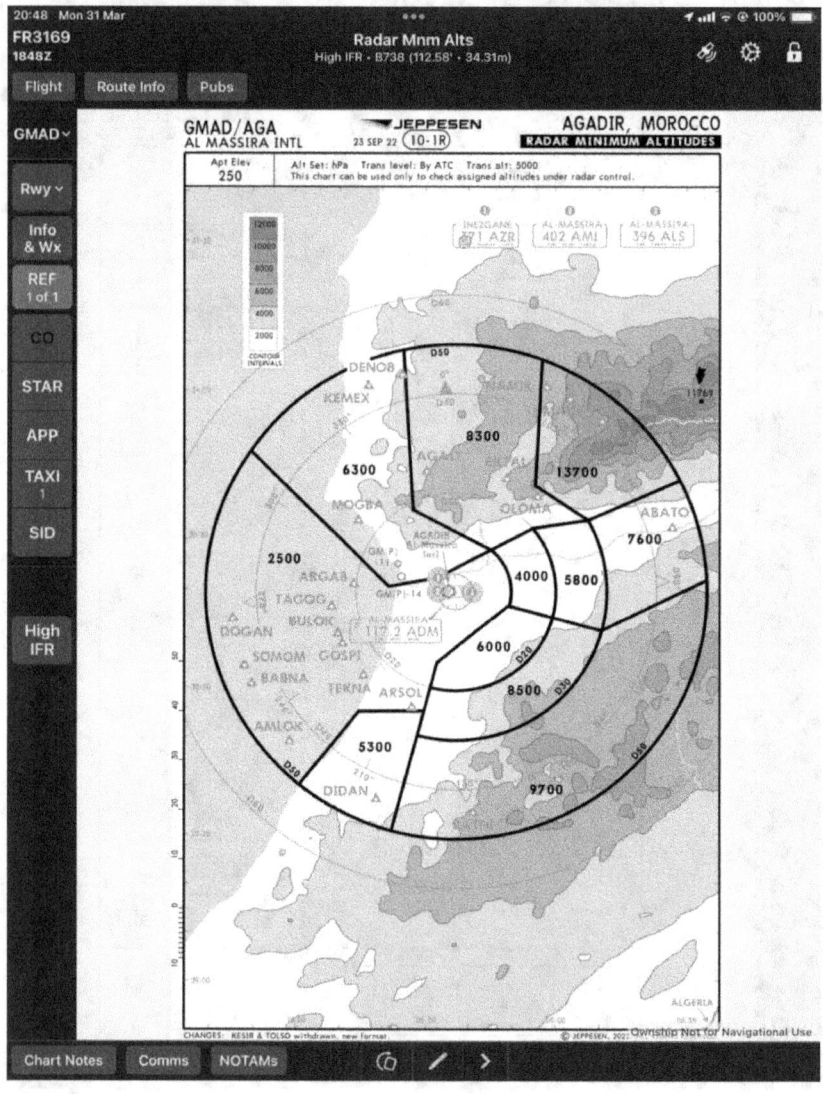

AMSTERDAM ILS & LOC APPROACH RUNWAY 36R

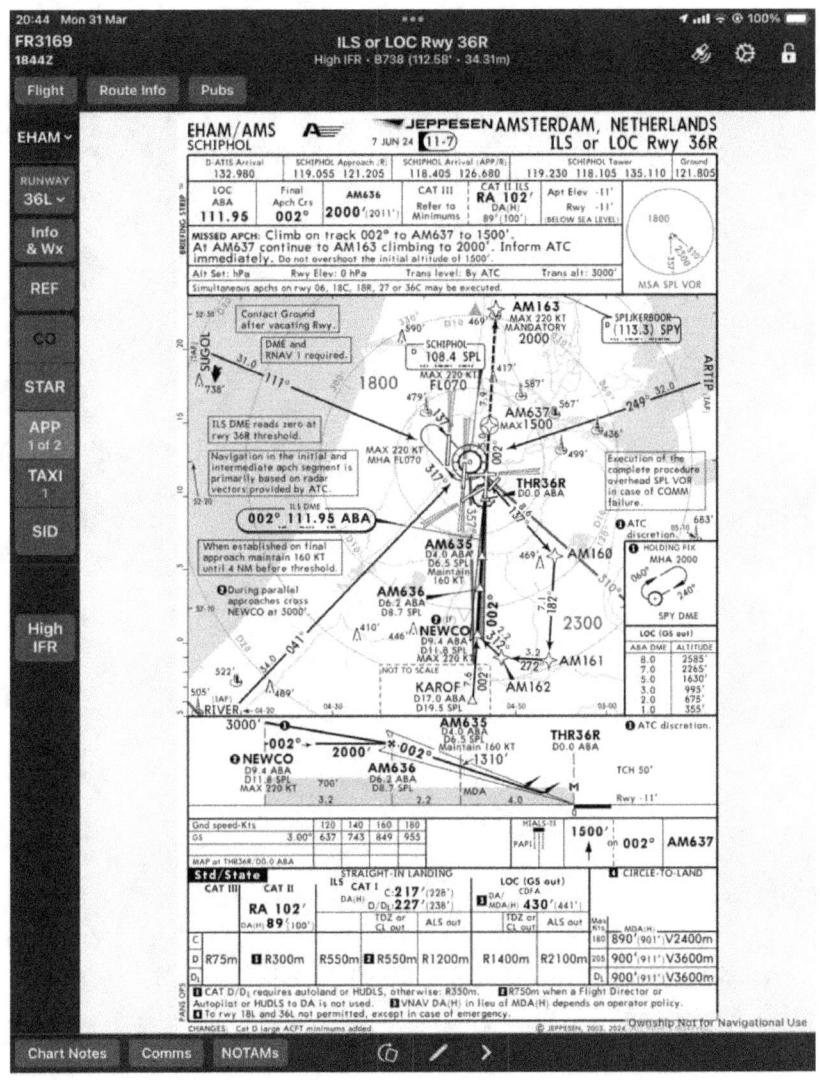

www.ingramcontent.com/pod-product-compliance
Lightning Source LLC
Chambersburg PA
CBHW070623220526
45466CB00001B/84